T0305203

MANAGING PEOPLE (INCLUDING YOURSELF) FOR PROJECT SUCCESS

Gordon Culp
Anne Smith

JOHN WILEY & SONS, INC.

New York / Chichester / Weinheim / Brisbane / Singapore / Toronto

Van Nostrand Reinhold   International Thomson Publishing GmbH
115 Fifth Avenue   Königswinterer Str. 418
New York, NY 10003   53227 Bonn
                                        Germany

International Thomson Publishing   International Thomson Publishing Asia
Berkshire House,168-173   221 Henderson Bldg. #05-10
High Holborn, London WC1V 7AA   Singapore 0315
England

Thomas Nelson Australia   International Thomson Publishing Japan
102 Dodds Street   Kyowa Building, 3F
South Melbourne 3205   2-2-1 Hirakawacho
Victoria, Australia   Chiyoda-Ku, Tokyo 102
                                        Japan

Nelson Canada
1120 Birchmount Road
Scarborough, Ontario
M1K 5G4, Canada

BBR   16 15 14 13 12 11 10 9 8 7 6 5 4 3

Library of Congress Cataloging-in-Publication Data
Culp, Gordon L.
     Managing people (including yourself) for project success / Gordon L. Culp, R. Anne Smith.
          p.   cm.
     Includes bibliographical references and index.
     ISBN 978-0-4712-9018-6
     1. Industrial project management. I. Smith, R. Anne. II. Title.
HD69.P75C85   1991
658.4'04—dc20   91-39039
                                                             CIP

Contents

Preface

Project success: every project manager wants it, but not every one gets it. Even when diligent about thorough project plans, budgets, schedules, progress reports, and so forth, consistent project success proves elusive. Why?

Why are some coaches consistent winners while some are not? Surely every football coach in a professional league knows the mechanics of blocking and tackling, and has a good repertoire of plays. Most coaches probably know the mechanics as well as the legendary Vince Lombardi. Yet Lombardi is a legend because of his consistent success, even though his players were perhaps not always the best individuals in the league. He had an exceptional ability to focus his substantial energy on each player as a person and to know how to motivate each one to his limit.

Have you ever gone to concerts by well-known and successful singers? All have an excellent grasp of the mechanics of singing. Some provide a stimulating, exciting evening of entertainment. You enjoyed them in person far more than from their records. You felt a personal connection. Others leave you feeling disappointed. You would rather have stayed home and listened to their records. The excitement and stimulation comes from those that connect with the people. The disappointment comes from those that never step beyond the mechanics of performing to establish rapport with the audience—no matter how excellent the mechanics are.

Project success also depends on more than mastering the mechanics of management. There are a multitude of books and articles on tools that are useful in the mechanics of project management. The mechanics are easy to learn. The intricacies of working with people are not. People do projects; mechanical tools do not. Certainly, there is a threshold of technical competence that must be present. Then, consistent project success comes when project managers focus their energy on people—the ones on their project teams, others in their organizations, and their clients. This is a book about person-centered project management, not about the mechanics.

Between us, we have 40 years of experience in managing and doing projects. They've ranged in size from a few thousand dollars to a few million dollars. The more we do, the more we see the patterns—project success comes when technical skills are combined with effective people skills; people skills more often cause a project to fail than does the lack of technical skills. Our appreciation of the importance of a person-centered approach for project success came from working with a wide range of people on a lot of different projects. Our curiosity about which personal attributes caused some project managers and teams to be so much more effective led us both to obtain graduate degrees in applied psychology, to supplement our engineering degrees and technical experience.

This book integrates our experience in the technical and humanistic aspects of project management into a practical guide to creating project success. We are two practical, practicing project managers, and we anticipate that this book will appeal to others wanting practical approaches to real world projects. These approaches will apply to any size or type of project. Life itself is a "project" to which many of these approaches will apply.

The formal training education that many project managers have includes little, if any, information on the approaches in this book. Other books unfortunately deal with the mechanical aspects of project management or often theoretical aspects of psychology, without integrating the two from a practical perspective. Our goal is to achieve this integration.

We owe thanks to many who have assisted us. Laurie Smedley Crist devoted many hours to preparing the manuscript, offering comments, and redrafting. Elizabeth Moattar assisted in preparing the figures. Several coworkers and friends offered comments, including Bud Benjes, Dick Bell, Mary Wees, Russell Culp, Jim Peterson, and Janet Dossigny. We received support and encouragement from Greta Breedlove and Carl Monahan. Drs. Ron and Mary Hulnick of the University of Santa Monica provided inspiration as well as insightful training in applied psychology. We've built upon the work of many who have preceded us in various aspects, as noted in the references for each chapter.

We hope that you will find this book a useful starting point on a life-long journey of continuous improvement in creating success and enjoyment in your life's projects, whatever they may be.

1

The Person-Centered Approach

STRUCTURE OR PERSON-CENTERED APPROACH?

A project is carried out by an *organization of people* brought together to accomplish a specific objective. In organizing and managing the project, you can emphasize the *organization* (structural) or you can emphasize the *people*. You cannot afford to ignore either. Often, there is great emphasis on the structure, with little on the people. This probably occurs because project managers are trained to use structural management tools but have little training to deal with people. Teachers like to focus on the tools, because they are easy to teach and make it easy to test the learning. Many benefits can be obtained by shifting the emphasis to the people aspects—the basis for the person-centered approach to project management.

So much of what we call management consists in making it difficult for people to work.

Peter Drucker

In the structural approach, there is centralized control of individuals and groups comprising the project team. Stability and predictability from project to project are valued and measurement and documentation are emphasized, with well-defined rules and procedures. Deviations from standard procedures must be approved by a central authority. Managers focus on coordination and on seeing that standard procedures are used. Everyone on the project team has a detailed description of their role, and each individual or group has detailed limits on the activities they are allowed to do. Management clearly delineates what resources each group has and what it does with those resources. This approach offers superficial comfort to upper management because it gives the appearance of a very high degree of control. With such an approach, the project manager seems to have tight control of the budget and schedule. The flow of information is also controlled and limited. Project management has the last say. Oftentimes, organizations favoring such an approach develop detailed strategic and tactical plans and organizational structures two to three years in advance. They produce planning reports replete with beautiful charts full of boxes and lines. There is little upward communication and very little, if any, participative decision making. Unfortunately, such an approach overlooks the fact that changes in an organization and its approach to project management must grow in response to the realities of the day-to-day working world and that the best information comes from the people on the front lines of the business—the working staff. Detailed, centralized planning with tight controls kills creativity.

Treat people as if they were what they ought to be and you help them to become what they are capable of being.

Johann W. vonGoethe

In the person-centered approach, there is a clear definition of the project mission. There is not a rigid definition dictated by the project manager of how each individual or group will pursue and conduct work within its general area of activity. Flexibility, creativity, and spontaneity are encouraged. Rather than attempting to control people, the emphasis is on inspiring them. Project managers develop progressive ways of meeting opportunities. There is an open, effective flow of information and participation in decision making. The project manager views the mission as more than technical or bottom line oriented; it's also viewed as a chance to coach and develop the team.

Ouchi points out that a basketball team offers an analogy to illustrate the difference in two approaches to managing project teams. The basketball team faces complex problems that occur at a high rate of speed. An effective team solves these problems without formal relationships and a minimum of individual specialization. The players understand their own tasks, skills, and their relationship to other tasks and players so well that coordination seems to occur naturally. The coach provides the training they need to hone their skills, a clear sense that success is measured in terms of team—not individual—results, and an environment that encourages each player's growth and contribution.

Compare this successful team to one where the coach attempts to control the players with a highly structured approach. The coach forces each player to stick to a strict role and to carry out specific, prescribed plays and moves. In an extremely strict, structured approach, the players might be told that the guards will always bring the ball down the court, the forwards will shoot, and the center will rebound. It is not hard to see the disastrous results that would occur if the forward couldn't rebound even if the ball fell into his or her hands. Such a heavily structured team will never perform as effectively or gracefully as the team where each member has the freedom and is encouraged to apply his or her strengths to each situation as it arises. The successful coach establishes an environment where using individual skills for team success rather than for individual goals (such as the highest individual scoring average) is rewarded. The Boston Celtics professional basketball team won 16 championships without ever having the league's leading scorer.

In any basketball team, project team, or organization in which the project is carried out, there will be a mix of the structure- and person-centered approaches; however, the differences in results depend on which predominates. As the leader, you have a major effect on which it is. Basically, you will either be inclined to want to hold control close to yourself, demonstrating little faith in the staff (structure-centered) or you will have faith in the judgment and abilities of the staff (person-centered). This doesn't mean that you don't have a system to monitor and course correct (Chapter 8), but the emphasis is on the people, not the structure. Because of the impact of the leader, we've devoted a separate chapter to leadership and motivation (Chapter 3). This chapter discusses the advantages and characteristics of the person-centered approach.

THE PERSON-CENTERED APPROACH

We're going to be so short of workers by the end of the decade, particularly knowledge workers and information-service workers, that companies that try to manage in the old top-down, hierarchal, drill-sergeant way are just doomed.

James A. Autry

The person-centered approach will motivate and retain key people to a far greater degree than a rigidly structured one. Undue constraints will discourage and drive away the creative, high-energy people most needed to create successful projects and will retain the ones needed least—mediocre performers comfortable only in a rigid structure. The person-centered approach demonstrates an essential trust in the capacity of others to think for themselves. A typical justification for the structural orientation is that it is necessary to keep the staff under control, avoid budget overruns, minimize duplication of equipment, correct staffing mistakes, and so on. Using a rigid structure to control people who don't think or act responsibly is a mistake. Train or replace them—don't restrict those who can think, take initiative, and perform responsibly.

Common sense is . . . the master workman.

Peter Mere Latham

The person-centered approach credits staff with common sense and intelligence. It recognizes that *people* make the difference between project success and failure. Information passes freely to and from team members, as well as between teams. The responsibility for decisions on how to reach project goals is shared. Decisions on who works on what are made within each group. Because the means to reach project goals are self-chosen, individuals have ownership and will invest their passion as well as their intellect in achieving the goals. A project manager provides resources and the environment that tells the staff that their intelligence and judgment are trusted. Only their own creativity, skills, and energy limit what they can do. The project manager does not draw a box around them. They are cared about as individuals, their input is valued, and management openly listens to their ideas. The decisions that affect their careers are in their own hands. They are the ones most affected by these decisions that they make. People become more accountable and responsible for their own growth and advancement, rather than placing that responsibility on the system or organization. With fewer constraints, people work harder, have more enthusiasm, are more creative, and grow more quickly. They motivate themselves. The person-centered approach is also more responsive to changing conditions because those closest to the work are making the decisions.

A flexible, person-centered approach fuels the ingenuity, creativity, and ambition of its people—the resource that fuels the success of any project and organization. Indeed, this approach underlies the modern quality management concepts that have propelled the Japanese to the forefront of quality products and services (see Chapter 9).

If a little knowledge is dangerous, where is the man who has so much as to be out of danger?

                                                                Thomas Henry Huxley

When the tasks at hand are complex, specialized, and changing, it is almost impossible for the project manager to know all of the relevant information. Higher quality solutions result from involved people. Not only are people more motivated and committed, the amount of information applied to any issue is greatly enhanced. Also, these solutions will be more effectively implemented when all team members work together to carry them out—and this is much more likely when all have been involved in developing the solutions. A project team usually can't function efficiently if everyone sticks strictly to a rigid job description and to the formal chain of command. So, the staff on every successful project has an informal network for working together without going through management. This network exists whether it is addressed or not. The person-centered approach recognizes the value of this informal network, influences it, and incorporates it into the project organization, thus increasing its effectiveness.

In the structure-centered approach, people tend to sit by and wait for their chance to argue their case for resources with the project manager. In a person-centered approach, the people work directly with each other to resolve problems and allocate resources. Individual strengths are recognized and supported rather than perceived as competitive. Since the staff takes more responsibility, the manager has more time and energy to deal with the external environment and take on other tasks.

As the manager in the person-centered approach, you do not simply throw all decisions to the staff and sit on the sidelines. Most importantly, you establish and maintain an environment where individuals can flourish by giving effective feedback on performance, maintaining policies that provide the needed freedom, providing ready access to training and specialized consultants, and giving them the tools and freedom for easy and full communication. You get involved in the process and retain some decisions because of your expertise. By allowing the staff to work through and resolve as many issues as possible, you encourage excellence. Otherwise, the commitment and willingness for the staff to openly share information and honestly deal with each other will be lost. The manager is actively involved in developing staff capabilities to solve problems and make decisions as a group. When people ask for a solution to a problem, your first response is, "What do you think?" "What options have you come up with?"

No snowflake in an avalanche ever feels responsible.

Stanislaw Jerzy Lee

Give people responsibility for the decisions affecting their daily lives and it will enhance their ability to perform, as well as their sense of well being. Residents of a Connecticut nursing home were given seemingly trivial environmental decisions to make, such as when to see a movie or how to arrange their rooms. Not only did health and general psychological well being improve, *death rates* were 50 percent lower than a comparison group who was told that management was responsible for their well being.

A classic study reported in 1948 by Coch and French also demonstrates the power of involving people in changes. Workers in a clothing factory were divided into two groups. A change in procedures was put into place by two methods. In the first group, the change was explained to the workers and they were sent back to the job with new instructions. In the second group, the need for a change was explained and the group then discussed options for new procedures and agreed upon the change to be made. The results were strikingly different. In the first group, output dropped by one-third and 17 percent of the group quit within 40 days. In the second group, productivity rapidly increased above the preceding rate and no one quit.

The highest and best form of efficiency is the spontaneous cooperation of a free people.

Woodrow Wilson

Orsburn, et al., summarize the striking results recently achieved by companies who have adopted programs that give more power, responsibility, and freedom to their staffs—key elements of the person-centered approach to project management:

- Xerox Corporation and Proctor and Gamble reported a 30-percent productivity improvement.
- Federal Express cut service glitches, such as incorrect bills and lost packages, by 13 percent in one year.
- Cadillac Motor Car Division of General Motors reported a 71-percent drop in reliability problems over a three-year period.
- Shenandoah Life processes 50 percent more applications and service requests with 10-percent fewer people.

Participative management is not a threat . . . it is simply common sense.

Dr. Delmar "Dutch" Lander

A tremendous amount of energy can be released by getting people involved and by giving people power. Once you establish a participative project approach, it will be self-sustaining because it will appeal to the basic values of your team members. The more that everyone on the project team feels a sense of power and involvement, the more ownership they feel in the success of the project. The person-centered approach shares power, which results in job satisfaction and higher performance by:

- Making certain that people have the skills and knowledge to make good judgments.
- Seeing that tasks are accomplished in a way that develops staff capabilities.
- Keeping people informed by a free flow of communication in all directions.
- Delegating tasks that are critical to project success.
- Demonstrating the ability at all levels to understand other people's perspective (empathy) on personal as well as professional issues.
- Involving people throughout the project team in important decisions.
- Stimulating creativity.
- Acknowledging and giving credit for people's contributions.

CHANGING THE APPROACH

. . . companies are moving from the hierarchial and bureaucratic model of organizations that has characterized corporations since World War II to . . .[the idea that] what has to be done governs who works with whom and who leads.

Michael Beer, Russell Eisenstat, and Bert Spector
Harvard Business Review
November-December 1990

*"For most (U.S.) companies, nothing short of a philosophical break with the
past will suffice.*

The Wall Street Journal
May 2, 1989

An increasing number of American companies are changing their heavily structured approach to one with increased involvement by the employees. A few success
stories and important leadership and communication skills needed are discussed in
later chapters.

The typical starting point finds a project manager who sees that his or her job
is to create the plan for the project and motivate his or her staff to carry it out. As
Ralph Stayer (president of a company, Johnsonville Foods, that has successfully
made the change) observes, the expectation of the leader is that the staff will follow
you much as a herd of buffalo follow their leader. Unfortunately, as he notes, this
leadership model didn't work well for the buffalo—nor does it in modern times.
Buffalo hunters used to slaughter the herd by killing the leader. Once the leader
was dead, the rest of the herd stood around waiting for instructions that never came,
while the hunters killed them one by one.

When you decide to increase the involvement of teams on your projects, it isn't
effective to command them to take increased responsibility. You can't jam responsibility down people's throats. Plus, if they've never had it before, they may not
know where to start. Don't set your people up for failure. The state of shared
responsibility is not the goal in itself, but rather the creation of an environment
where people insist on being responsible.

Change the environment; do not try to change the man.

Buckminster Fuller

The most effective way to change behavior is to put people in a new environment that fosters new roles, responsibilities, and relationships. Your power as a
leader lies in your ability to manage the context of the project organization so that
it shapes the way people think and what they expect. You can encourage people to
be responsible for their own performance and to own their own problems. You can
plan individual changes, but you can't plan or control the whole process of change.
You can learn that expectations often become reality and that you can send
messages that will shape expectations to the project's advantage, and you can
learn that encouraging continual change is your real job. You learn that you must
develop a group that is strong enough to let you wade in with opinions, but without
fear that you'll inhibit the discussion.

If you're facing the prospect of changing the project environment from a highly
structured approach, you're probably asking yourself how the change will be
received and how you can delegate responsibility when you're not sure that your

team wants it or is capable of acting responsibly. There are several steps typically involved in changing your approach. Unless you are involved in an unusually long-term, very large project, you are not likely to accomplish all of these steps in a single project. Rather, there are steps to be taken over time to alter the approach taken to projects within your organization. We will discuss these steps briefly here. We will devote the rest of this book to the skills that are involved in both the transition to and use of a person-centered approach.

It is the circumstances and proper timing that give an action its character and make it either good or bad.

Agesilaus (400 B.C.)

First, you must assess whether or not the time and environment are appropriate for the change. If for some reason you're faced with a staff that is already working at the limit of its abilities, it would be counterproductive to ask them to take on more responsibilities until you have added people with the needed skills or aptitude to learn new skills. Timing can be a factor as well. In one case, a change was postponed because several people had been recently laid off. The remaining staff needed time to adjust before it was prudent to ask them to take on more.

Nothing will change until you take action to change yourself.

Ra Bonewitz

The chief factors affecting timing are *your* willingness and commitment to change. Remember, responsible action by others does not mean that they will come up with the same answers that you would. If you're equating independent decisions with their ability to reach the decision you would have, then *you* are not ready for the change. You need to first focus on changing yourself before you begin to change the staff. Your own readiness is important because false starts down a path of change can be disastrous. Those wanting a change for excellence will be crushed. The cynics who wrote it off as a passing fad and were fearful of change will be reinforced. Both reactions will make it even more difficult to implement change in the future.

Don't be afraid to take a big step if one is indicated. You can't cross a chasm in two small jumps.

David Lloyd George

Don't be fearful of making mistakes. If you hesitate, people will take that to be your unwillingness to change.

Secondly, you need to optimize the environment for change. It is not likely that you'll ever find an absolutely perfect set of conditions for change. For example, some of the staff may be ready and able to take on more responsibility while others are not. In the first step, you determined that the environment was suitable for a change. Now it is time to change it. Such steps as restructuring the staff, changing assignments or procedures, instituting training, or establishing groups to solve coordination problems may be useful. Involving staff in these steps can also be excellent preparation for change. Some of the most effective results are obtained when you allow the staff to start the process of change without specifying an approach.

Next, it is important to assess who outside your project may be affected by the change so that you can lay the groundwork with them. For example, to avoid surprises, are there others in your company that need to know of the changes you are making? They need to understand the new responsibilities in order to make sense of communications with your project team. Certainly, if you report to another person, getting your boss involved in the process is essential. Your boss, when properly involved, will be supportive when it is clear that your goal is to improve the effectiveness of current and future project teams.

It is only the wisest and the very stupidest who cannot change.

Confucius

The fourth step is to work one-on-one with key staff to establish credibility that things are changing. Informing key staff of the change before group presentations reduces the potential for a negative group reaction. They become your ally in group meetings, broadening your base of credibility. One approach that demonstrates your readiness to change is to ask the staff what you could do to improve your management skills or style. Such an approach not only demonstrates your willingness to change, but may also improve previously antagonistic relationships. However you do it, it is critical that you are truly open to input during the process of making change. There is a danger that you'll get so involved in your own ideas that you will not be receptive to input. Key individuals will sense this and feel closed out of the process. Making these key people participants in the process of change is important in gaining their commitment. By building solid understandings with individual key staff members, a solid foundation for building a team is established. Having them on your side helps immensely. In addition to your readiness to change your own approach, your key staff need to be ready to change, to be open to more involvement by the teams they lead. You can assist the process by encouraging them to do some reading on the person-centered approach and to take training in leadership and interpersonal skills. We have more to say about this in Chapters 2 and 3.

The fifth step is to clearly communicate the change to your staff, using terms that everyone can understand. A significant and unexplained change in your management style will cause a great deal of confusion and wasted energy on the part of your staff as they try to figure out what is happening. State clearly what you are trying to do. It will take more than words because your staff includes cynics who are sure this is nothing more than a passing fancy. Ask them to withhold judgment even if they don't feel they can be supporters. Ask them to check it out and give it a chance. Before they become committed to the change, they have to believe that it is permanent. Your actions must match your words. People will accept and even seek change when it helps them grow and expand. Emphasize the increased individual opportunities they will have to make a difference. Set up a vehicle for discussing ideas and problems with those involved in the change.

By the mid-1990's, we'll define good management as the ability to get out of the way.

David Luther

The successful change at Johnsonville Foods, described by Ralph Stayer, was accompanied by such concrete actions as allowing the employees to determine how to divide up the company profit-sharing bonuses; allowing departmental teams to hire, train, evaluate, and fire fellow employees; giving each person a $100 educational allowance to spend on any developmental activity he or she desired; and Stayer removing himself from meetings so that others could make their own decisions. When the leader starts the process of change, the staff is much more likely to follow. After all, it is hardly fair for you to take the position that you are committed to change but that your staff must go first. Even relatively small actions demonstrating change on your part can be effective when repeated over a long period of time.

What a bore it is, waking up in the morning always the same person.

Logan Pearsall Smith

The sixth step is to change your meetings from sessions where information is exchanged or transmitted to sessions where significant issues are resolved through the participation of all involved. An effective method is to use project team meetings to deal with increasingly important issues and delegating the decisions to the team. Dealing with real issues reduces resistance to change. For example, having the team make the decision on project budget allocation and individual assignments will quickly get everyone participating. A key here is to have the meeting become the team's, not yours. Get team input on the agenda. Rotate the meeting chairmanship among group members. Make sure that discussion occurs among team members equally and not between individuals and you. The successful transition may take more than just altering your behavior at meetings. You can try insisting that everyone speak their mind only to find that their behavior is still influenced by the way you ask questions, the tone of your voice, or your body language. Even staying silent at the meeting may not work because others won't commit themselves until they get a reading on your position. You may have to simply not attend the meetings to force the team to accept the problems as their own to be solved by themselves.

And often enough our faith beforehand in an uncertified result is the only thing that makes the result come true.

William James

You do face some risks as developing teams resolve issues. Some problems may be created, but foot dragging can create missed opportunities. Discussing resulting problems openly in the team is a strong developmental tool. As they realize they must live with the consequences of their decisions, an inherent quality control mechanism surfaces. Many individuals will quickly embrace the spirit of greater responsibility and will take the initiative in proposing and developing solutions or new programs. By keeping everyone involved in a team, those members slower to accept the change will see what the others are accomplishing and will become more receptive. Your role changes in the process from that of a problem solver to a builder of the team's capacity to solve problems and to providing the right environment for change and growth to prosper. You must stay engaged and not withdraw—but you are engaged in building the team's capacities and willingness to share responsibility, rather than in personally solving the problems. Your initiative puts the team into motion. Your trust in their inherent desire to do good work keeps them in motion.

Caring is contagious. Help spread it around.

Harvey Mackay

Throughout the above steps, you should be developing individuals. You can support development by defining the skills and attitude to be learned, including time frames and expected results. Often, in the past, individual development has meant sending people to training sessions or giving them new assignments. The most effective development occurs through on-the-job learning combined with coaching, counseling, and training courses. Recognition by you of personal improvements provides motivation to keep learning and growing. A continuous process through challenging assignments consistent with individual abilities and interests, combined with coaching and effective feedback on performance, leads to effective development. You should continually be looking for ways to stretch individual competencies while getting the job done. You set higher standards while providing more assistance and support. Having high expectations of individual abilities will create improved performance. Open communication is essential because the individual needs to feel open (and safe) to ask for more help or say when he or she is getting too much help.

Another important step is to involve the team in developing a clear mission statement for the project. A mission statement basically addresses the question,

"What are we here to do?" Here's an example statement: "This project team exists to produce plans and specifications by July 1, 1994, for a wastewater treatment plant for the city of Portland that meets the needs of the city in the most cost-effective and reliable manner possible." Allowing the team to develop the mission statement may reveal, in some cases, a lack of agreement about the project mission. The resolution of these differences can be a powerful learning and growth experience. A strong, common point of view will result. Defining a clear, concise statement of the project mission can be a struggle, but it can be a struggle that produces much learning, commitment, and team building. The project mission statement should:

- Support the mission of your client. In the sample, the client's mission was cost-effective, reliable wastewater treatment.
- State why the project exists and its major activities.
- Be concise and unambiguous. Use words that describe the mission precisely.
- Be succinct so that it can be easily remembered and used.
- Clearly define the limits of the project. The sample makes it clear that this team is not doing process research or construction—it is doing design.

Generating the project mission statement creates a framework to generate goals and schedules. It shows each team member how their work supports the project.

We've presented some general guidance on the processes involved, but there is no universal formula for the transition from the heavily structured approach to one that is person-centered. We've talked about the process of changing in terms of steps. Unfortunately, the real world is not so neatly ordered and you'll probably find that several steps will overlap. Indeed, the steps often interact so as to reinforce each other. You should view them as part of a continuous, flowing process rather than any kind of rigid, step-by-step procedure. Underlying a successful process will be your adopting a role of developing conditions that allow your staff to continually grow into increased responsibilities and to stimulate continuous change. You look at the problems that arise as opportunities that can be used to allow individual growth.

Effort supposes resistance.

                                                                Charles Sanders Pierce

You may encounter resistance as you change to a person-centered approach. Resist the temptation to interpret it as foot-dragging. Recognize that resistance to change is a bit like pain—it tells you something is wrong, but not what is wrong. When resistance appears, you'll need to take the time to listen carefully to find the cause of the trouble. Reservations must be openly discussed before changes can occur. Often, it isn't the technical aspects of the changes that are the problem, but rather that the change is threatening established interpersonal relationships for doing work. Open and candid discussion of concerns will do much to establish a sense of trust. You'll need to develop and apply the skills described in several of the subsequent chapters of this book—effective communication, motivation and leadership, increasing your personal effectiveness, and integrating these skills into an approach to produce high-quality work. Many of these same basic skills also directly apply to areas essential to successful person-centered management—negotiations, productive meetings, and planning and monitoring projects—that are addressed in the remaining chapters of this book. We suggest that you read the entire book, reflect upon your approach to project management, and make the changes that you're enthused about. Once you embark on the process toward a goal of person-centered management, you'll probably find that your goals keep changing because the process is a journey, not a destination. Bon voyage! We'd love to hear about your trip!

When you're through changing, you're through.

Bruce Barton

References

1. Rogers, C.A. 1980. *A Way of Being.* Boston: Houghton Mifflin Co.
2. Kouzes, J.M., and B.Z. Posner. 1987. *The Leadership Challenge, How to Get Extraordinary Things Done in Organizations.* San Francisco: Josey-Bass.
3. Stayer, R. 1990. How I learned to let my workers lead. *Harvard Business Review,* p. 66. November–December 1990.
4. Bradford, D.L., and A.R. Cohen. 1984. *Managing for Excellence.* John Wiley & Sons.
5. Coch, L., and J.R.P. French, Jr. 1948. Overcoming resistance to change. *Human Relations,* p. 512.
6. Laurence, P.R. 1954. How to deal with resistance to change. *Harvard Business Review,* May–June 1954.
7. Baer, M., R.A. Eisenstat, and B. Spector. 1990. Why change programs don't produce change. *Harvard Business Review,* p. 158. November–December 1990.
8. Orsburn, J.D., L. Moran, E. Musselwhite, and J.N. Zenger. 1990. *Self-Directed Work Teams—The New American Challenge.* Homewood, Illinois: Business One Irwin.
9. Todryk, L. 1990. The project manager as team builder: creating an effective team. *Project Management Journal,* p. 17. December 1990.
10. Ouchi, W.G. 1981. *Theory Z. How American Business Can Meet the Japanese Challenge.* Reading, Massachusetts: Addison-Wesley Publishing Co.

2
Communication

What we've got here is failure to communicate.

Prison warden to Paul Newman in movie
"Cool Hand Luke"

People often blame their problems on a failure to communicate. Our guess is that it is the most popular choice of causes-to-blame. This popularity is not without reason. Management consultant Peter Drucker claims that 60 percent of all management problems are a result of faulty communications. Misunderstandings lead to problems on projects as well as in life.

The need to communicate goes far beyond the fact that good communication will improve your performance as a project manager. It is a basic human need. Prisoners of war have reported that prisoners that couldn't communicate died much sooner. Prisoners often risked death to communicate by writing under plates, coughing, singing, tapping on walls, or flapping laundry a certain number of times to transmit a letter of the alphabet. Captors have often taken great pains to prevent communication. They knew that prisoners could stand more pain if they could communicate with others.

Several surveys have found that the ability to communicate well is ranked as the most important key to success by leaders in business and the professions. Unfortunately, communication is a skill that many project managers receive no formal education in and one that they don't cultivate. Although this is not normally a do-or-die situation for a project manager as it was for the POWs, effective communication is the critical link to your project team. It is a vital part of the project management elements discussed in subsequent chapters.

The most immutable barrier in nature is between one man's thoughts and another's.

William James

Unless the thought barrier between you and your project team is penetrated, you simply cannot function as a project manager. Communication is your sole tool to break down this barrier.

"BENEFITS" OF POOR COMMUNICATION

I have suffered from being misunderstood, but I would have suffered a hell of a lot more if I had been understood.

Clarence Darrow

Graham has identified several "benefits" of poor communication:

- Minimizes impact of poor planning (don't let others know you do not know what you are doing).
- Easier to deny what you said later on; preserves freedom to change your mind.
- Good techniques to mask your true intent.
- Allows you to say two things at one time.
- Avoids the need to share credit for your idea.
- Helps you preserve mystique and insecurities.
- Helps minimize opposition and criticism.

To these, we can add the encouragement of procrastination and of superficial planning. In the environment of project management, these "benefits" have a negative value. This chapter will give you the tools to eradicate poor communication. First, we'll discuss basic communication skills and summarize the most important tools for communications. The chapter concludes with applications of these skills to the presentations you make to sell or explain your project to others.

BASIC COMMUNICATION SKILLS

The two words "information" and "communication" are often used interchangeably, but they signify quite different things. Information is giving out; communication is getting through.

Sydney J. Harris

Active Listening

A good listener is not only popular everywhere, but after a while he knows something.

Wilson Mizner

Man's inability to communicate is a result of his failure to listen effectively, skillfully, and with understanding to another person.

Carl Rogers

It's been said that we've been provided with two ears and one mouth so that we could listen twice as much as we talk. Unfortunately, few people do. Most people love to talk and hate to listen. Talk is cheap because supply exceeds demand. Each day we encounter many examples of poor listeners. They are so busy working their mouths that their ears have little chance to function. According to one statistician, the average person spends 13 years of their life talking. Each year, their spoken words would fill 66 books, each 800 pages long.

We've begun this section with listening because it is the single most important attribute of an effective project manager. This conclusion is supported by a Loyola University study of hundreds of businesses. They found that a manager's ability to truly listen to his staff was the most important.

Listening is not merely not talking, though even that is beyond most of our powers; it means taking a vigorous, human interest in what is being told us.

Alice Duer Miller

Many people take listening for granted. They confuse hearing with listening. You may hear every word but if you're not listening, you'll misunderstand the message. Studies have shown that the average listener typically comprehends about half of what is said. Within 48 hours, the retention drops to about 25 percent. After a week, only 10 percent or less is retained. Is it any wonder that projects so often get sidetracked by poor communication?

Why is it that listening is so difficult? One reason is a natural tendency to immediately evaluate or judge what the other person says without truly listening. For example, if someone were to say to you, "Richard Nixon was the worst president in U.S. history," your immediate response will likely be either approval or disapproval. You will evaluate what was said from your frame of reference. You may also be making judgments, such as the speaker must be a Democrat, a liberal, or a poor (or good) thinker. This tendency to evaluate is even stronger when emotions are involved. If a person says, "Your work is really poor," your tendency to evaluate and judge will be even greater. There will be little or no communication.

Active listening occurs when you avoid this tendency to judge and evaluate. You listen with understanding. You see the other people's point of view, sense how it feels to them, and experience their frame of reference. This is called empathy. Empathy is not necessarily agreeing with the other person, but is recognizing and understanding their point of view. It is different from sympathy. It is not feeling sorry or pity for them. It is putting yourself in their shoes and feeling what they are feeling. It might take the form of a statement, such as, "I'm hearing that you feel you are not appreciated. I understand why you might feel that way. If I were in that situation, I'd feel the same way."

Carl Rogers' pioneering work in psychology has shown that truly empathic listening—feeling the issue as the other person feels it, understanding and sensing the emotional flavor it has for the other person—releases powerful forces of communication.

You can listen like a blank wall or like a splendid auditorium where every sound comes back fuller and richer.

Alice Duer Miller

There is a simple exercise that will show you the difference between good listening and active listening. In a discussion with a group or another person, adopt the rule that everyone can speak for themselves only after they have restated the ideas *and* feelings of the previous speaker to the speaker's satisfaction. If you want to interact effectively with people, you need to first understand them. You need to do more than just be able to feed back what they've said. You need to listen so that you can get inside other people's frame of reference and understand how they feel. Before you speak, you'll have to understand the speaker's thoughts and feelings so well that you can summarize them. Try it. You'll discover: 1) it is far more difficult than you can imagine; 2) your own views change; 3) emotion drops from the discussion; 4) differences are reduced; and 5) whether you agree with the other person or not, being able to accurately experience and restate the other person's point of view will establish *real* communication. When your intent is to understand rather than judge, distortions and miscommunication drop away. You'll find that communication becomes pointed toward solving a problem rather than toward attacking a person or group. When we each see how the problem appears to the other, amazing results can occur. The beauty of this skill of active listening is that you can initiate it without waiting for the other person to be ready.

Even if you are a good listener, you may not be enjoying the benefits you'd get from being a truly active listener. As only a good listener, you are able to repeat back most of the speaker's words, but you're concerned with the content rather than the feelings, ignoring the nonverbal messages from the speaker's body language, facial expressions, or changes in tone. You'll believe that you understand your project team or customer, but, to your surprise, you'll find that they don't feel understood. Because people can comprehend speech at a much faster rate (500 words per minute) than others can speak (125 words per minute), you find your mind racing ahead to anticipate the speaker's words and forming your own evaluation and response. This is *not* active listening, even though you thought you heard everything the speaker said.

Weigh the meaning and look not at the words.

Ben Johnson

In active listening, you refrain from evaluating speakers' words, and you see and feel things from their point of view. You are putting yourself in speakers' shoes, feeling their feelings. You listen for the intent and feeling as well as content. You don't interrupt the speaker, and you use questions to encourage the speaker to continue and to clarify, such as, "I want to be sure I understand you. Would you please say a bit more?" or "Can you give me an example of what you mean?"

You are careful to add nothing. You maintain eye contact. Even though you hear with your ears, people judge whether you're listening by looking at your eyes. Gentle, intermittent eye contact is a key point of being an active listener. Provide nonverbal indications that you're listening such as affirmative head nods and appropriate facial expressions.

Let's compare good listening and active listening so that the difference is clear. First, if you're practicing good listening, you are able to repeat most of the content back to the speaker. At least, you are listening to what is being said. In itself, it's not a useful tool because people will soon feel that you're merely mimicking them without understanding them. Listening applied to one of your team members might look like this:

"The client just called. It's the fifth time he's changed his mind. We'll never meet the deadline."
"So, the client just changed his mind again. You think we'll miss the deadline.

You've repeated back the content, showing you were listening. However, you haven't evaluated or interpreted their feelings, showing your understanding. Active listening would look like this:

"The client just called. It's the fifth time he's changed his mind. We'll never meet the deadline."
"It seems to me that you're really frustrated about the client's changes."

Frustration is the feeling and the client's changes is the content. When you feel the issue as other people do, you will experience some amazing results. By helping them work with their feelings and thoughts, they will accept that you really want to listen and understand them. You establish a direct link with what's going on inside of them. Their thoughts, feelings, and communications all become one. This skill works only if you *sincerely desire* to understand the other person. People will quickly see through any attempt to manipulate or control them using the skill only as a technique. Truly active, empathic listening will make others feel so safe that they will open up until they get to the core of the problem. Let's see where active listening might have taken the conversation:

> "The client just called. It's the fifth time he's changed his mind. We'll never meet the deadline."
>
> "It seems to me that you're really frustrated about the client's changes."
>
> "Yes. The project seems to be totally out of control. We don't know what the client wants."
>
> "So, do you feel that we haven't listened to the client?"
>
> "Well, partly, I suppose. Most of his changes seem to be things he's had on his mind for quite a while."
>
> "Do you feel he's had some ideas that we didn't hear?"
>
> "Right. He throws out so many ideas. I bet he's frustrated with us for not digging deep enough at the start. We'd better flush out his ideas and regroup."
>
> "So, I'm hearing that you would like for us to talk with the client and reassess where we're going on the project."
>
> "Right. If it's OK with you, I'd like to arrange a meeting."
>
> "Good idea."

By actively listening, you got through the surface feeling of frustration about the client to the core issue of frustration over the project team not listening carefully enough to the client before work started.

It's a luxury to be understood.

Ralph Waldo Emerson

You will find that others will feel good about you because you have made an effort to understand them from their point of view—and you'll learn plenty in the process.

If listening is such a powerful communication tool, why is it so poorly used? Poor motivation is probably the most common cause. People hear what they want to hear and shut out what they don't want to hear. Unless you enter a conversation with the empathic attitude of the active listener, you are not going to fully understand the other person. Another cause is lack of concentration. Active listening takes a concerted effort to concentrate, overlooking outside distractions.

No man would listen to you talk if he didn't know it was his turn next.

Edgar Watson Howe

Some people consider listening as a passive, powerless act. They equate speech with power. Actually, when you actively listen, you will hear others telling you how to best meet their needs. They will feel truly heard and will be much more attentive to you when you do speak. There is much power to be gained from listening. Another cause is the tendency to prejudge the speaker based on appearance, immediately distorting your perception of what is said. Setting aside such judgments is essential to active listening. Many feel that listening takes too much time. It may take more time initially, but will save a great deal of time later. Imagine the results if your doctor told you that he was too busy to make a diagnosis and immediately handed you a prescription. That's what many people do when they don't take the time to truly listen before offering their advice or opinion.

Here are a few basic rules for active listening:

• You can't talk and listen to others at the same time. This is a pretty basic concept, but one that is broken by many—probably most—people. People are so anxious to make their own point that they are formulating their response or even begin speaking it while the other person is still talking.

It's all right to hold a conversation, but you should let go of it now and then.

Richard Armour

• Be empathic. See the speakers' point of view. Set aside emotions and judgments. Put yourself in their shoes. Restate their comments. If you can't restate their ideas *and* feelings, you aren't listening. Don't let their appearance affect you.
• Listen for intent as well as content. The speaker's eyes, body, and face are all sending important parts of the communication. Integrate all of the information—verbal and nonverbal.
• Focus on the speaker. Relax and clear your mind. Shut out distractions. Don't fiddle with pens or papers or look out the window.
• Let other people speak first. They will reveal information that will help you address their concerns. Be patient. Allow them time to finish. Encourage them to clarify and elaborate. You can adjust your discussion after you have heard them out.
• Be attentive. Maintain eye contact—gentle, not staring. Don't cross your arms or legs—keep an "open" body position, leaning slightly forward.
• Ask open-ended questions (discussed in the next section) to encourage expression of feelings or thoughts.
• Use examples to clarify meaning. For example, your customer asks you to produce a report in the next three days while he is out of town. You need to be sure you know what he wants, to avoid an unpleasant surprise when he returns. You could show him an example report from a similar project or from another phase of his project. "Here's an example of a report that seems similar to what you're asking for. How close is it to what you're looking for?" The customer's response will help you quickly focus on what she or he's looking for.

By actively listening, you will encourage others to talk more openly about their needs, wants, goals, and suggestions. Once your project team (or customer) senses that you are truly listening and understanding them, they will be much more open to your point of view. People like to be heard.

A sensitive ability to hear, a deep satisfaction in being heard, an ability to be more real, which in turn brings forth more realness from others . . . therein my experience are the elements that make interpersonal communication enriching and enhancing.

Carl Rogers

Communication tools that are useful in active listening, such as asking questions, silence, and perception checking, are discussed in the next sections.

Questions

I kept six honest serving men / They taught me all I knew / Their names are What and Why and When / and How and Where and Who.

Rudyard Kipling

The ability to ask the right question at the right time is a powerful communication tool. Yet how many courses or seminars on asking questions have you attended or seen advertised? It is an often neglected aspect of communication.

The primary role of questioning is to open up communication. Once communication is flowing, you may shift the purpose of questioning to gain specific information, to check your understanding, or to learn more about the other person. Questions of your customer at the start of the project are important to understand his or her needs and desires. The ability of questions to create effective communication is their greatest strength. For example, one of the most effective methods of finding out the true status of work on your project is by asking team members questions. By asking questions and listening to and clarifying the answers, you will learn far more than you can from a written status report. Also, the other person will feel valued and will be more likely to listen to you. Effective use of questions will do much to increase: 1) your knowledge of the project; 2) your understanding of your customer's needs and desires; 3) cooperation among project team members; 4) team efficiency; and 5) project quality.

By using the right type of question and phrasing it properly, you will increase your chances of getting information, while simultaneously improving your relationship with the other person. There are two basic types of questions:

- Open
- Closed

Open questions cannot be answered by a simple "yes" or "no." They are very useful in gaining the involvement of the other person. They give other people permission to air their point of view in their way. They also demonstrate your respect for whatever content they want to share with you. They create an atmosphere of acceptance and interest rather than interrogation or an appearance of trying to lead the other person in a specific direction. They typically begin with "what," "when," "how."

Closed questions can be answered by "yes" or "no" or some other very brief answer. They are usually used to obtain specific facts. They do not generate much involvement because the answers often do not require much thought. Often, they are used to direct the conversation in a specific direction. They are the type of questions that most project managers are used to asking. In the rush to meet the project deadline, you may ask closed questions just to get the facts and get on with the project. By using open questions, you could learn much more about the factors underlying the performance of the project team members.

Compare the two types of questions and the responses you might get:

Closed	Open
Are you on schedule?	What are the factors that are having the most impact on your schedule?
Do you think we should do it differently?	What are the ways that you think we should handle this problem?
Do you agree with this approach?	How are you feeling about this approach?
Is that your approach?	Why do you feel that this is the best approach?
Do you really want to be reassigned?	What do you like least about your assignment?
Are things going OK?	How are things going for you?
Do you realize you missed the deadline again?	What problems are you having with the schedule that I could help with?

Open questions stimulate responses that go deeper into the other person's feelings, attitudes, and motivations. They can help the other person gain better insights into their own internal processes and come up with their own solutions to their questions or problems.

It is better to know some of the questions than all the answers.

James Thurber

One of the greatest temptations is to "fix" other people, that is, solve their problems for them by telling what to do to resolve their situation. The seeming advantages are that it would be a lot faster than listening to them thrash about, and, surely, since you're the project manager, you must know more about their problems than they do. If you're a technical expert in a given area, you may well be a valuable resource for them to get technical answers. However, in regard to management issues or counseling the project team on individual issues related to motivation, performance, or goals, trying to "fix" the other person is counterproductive. They have all the inner resources and information they need to deal with such personal issues. By careful use of open questions *and* active listening, you can be the catalyst to bring these resources to bear. There are *many* advantages to this approach. People will be fully committed to the solution since it is one they developed. It is the best solution for them. They will feel valued by you because you listened and assisted their problem solving process. They will be much more likely to listen to you. These are hardly new ideas. Lao-tse, a Chinese sage who lived 25 centuries ago, said:

It is as though he listened
and such listening as his enfolds us in a silence
in which at last we begin to hear
what we are meant to be. "

and . . .

"If I keep from meddling with people, they take care of themselves.
If I keep from commanding people, they behave themselves.
If I keep from preaching at people, they improve themselves.
If I keep from imposing on people, they become themselves. "

By using open questions and active listening, you'll find that people will move away from rigidity toward flexibility, away from dependence toward autonomy, away from defensiveness toward self-acceptance, and away from being predict-able toward creativity. In other words, the individual characteristics you need for a successful project team are stimulated.

The manner in which you ask questions can be as important as the content of the question. Ask permission to ask questions: "I'm not sure I'm following you. Can I ask you a couple of questions so that I can get a clearer understanding?" This act of courtesy will help put others at ease. Keep the language simple. Avoid jargon or using big words in hopes that they will be impressive. Keep the questions simple with only one major thought per question. The other person has to understand the question before you will get a meaningful response. Pursue one line of thought at a time. Questions should not jump from topic to topic. Use a relaxed tone of voice. Shouting a question at someone doesn't work. Disguising a statement as a question doesn't work: "Well, alternative A is the best, don't you agree." Your manner of questioning needs to convey the message that it is safe to respond openly, fully, and honestly. Allow plenty of time for the other person to respond. Be patient. Consider the personality and state of mind of the other person and adjust the timing and delivery of your question. *Actively listen* to the responses and adjust your questions accordingly.

The combination of active listening and effective questioning is a powerful communication tool to achieve project success. You will learn much from your customer and from your project team that will be essential to a quality result.

Silence

Who then tells a finer tale than any of us? Silence does.

Isak Dinesen

It may seem illogical, but silence can be a useful communication tool. Some of the most significant messages can be communicated without words. Leave the wrong thing unsaid. Simply maintaining your silence while you *actively* listen can give other people the freedom they need to fully express themselves. It can show that you care enough about them to truly listen. Silence can also provide you with useful information that simply wouldn't be gained if you were vying with other people for speaking time.

Many a time I have wanted to stop talking and find out what I really believed.

Walter Lipman

Perception Checking

To understand everything makes one tolerant.

Madame de Staël

We all have a tendency to hear what we want to hear. Unfortunately, this tendency is a barrier to communications. Frequently checking your perception of what the other person has said is the best method to break through this barrier. It is also important if the conversation is going to continue for any length of time and still have meaning. You can do this by restating in your own words what you heard other people say. The key is not to tell other people what *they* meant, but to only feed back what *you* heard them say. Perception checking seeks verification of the content and feeling of the other person's message. It will also often lead to clarification and elaboration by the other person. Be careful about the way you check your perceptions:

Effective	Ineffective
I heard you say . . . Is that accurate?	What you are telling me is . . .
Here is what I hear you saying.	From what you said, you must mean. . .
I heard that you want . . .	So, what you want is . . .
If I'm hearing you correctly, it appears your major concerns are . . . Is that correct?	I know exactly what you mean.

Stress what *you* heard, not what you are interpreting *them* to mean. By doing so, you'll transmit a message that they are important to you and that you want to do your best to understand them. The accuracy of the information you receive will also improve.

The great thing about human language is that it prevents us from sticking to the matter at hand.

Lewis Thomas

Perception checking is important even if you're hearing perfectly. You may hear every word, but may misunderstand how it is being used. The 500 most commonly used words in the English language have 14,000 definitions. For example, "post" can mean a pole in the ground, an office where a letter is mailed, a brand of cereal, the process or recording figures in a journal, the displaying of a public notice, informing someone, the stationing of a soldier or guard, a person's position with a company, and more. When someone says it is important to post last month's project costs, you should check your perception by asking something like, "I heard you say that we should enter last month's costs into our accounting system. Is that correct?" Maybe that's what they meant, or maybe they meant they should be mailed to the client.

Perception checking is not restricted to verbal content only. You should use it to check out the other person's feelings. For example, you could say something like, "You say you're satisfied with your progress, yet you seem really uneasy. I noticed you were wringing your hands and looking away. I sense that you feel differently. Is what I'm sensing accurate?" When body language doesn't match the content, it is especially important to check to their feelings. If someone scowls at you and growls, "This is going great," you should ask, "I sense you are upset. Am I correct?"

It is also important to check the perception that the other person has of what you said. You can ask questions such as, "What does that mean to you?" or "Would you mind summarizing the key points we've discussed?" "What's your understanding of what we've been discussing?"

Giving Feedback

It is important to give feedback to others, whether it be positive or corrective feedback. People need to know if they are on course or not. As a navigator uses compass readings, people must receive feedback to make course corrections or maintain their direction. People tend to be stingy with positive feedback. Done properly, both types of feedback are effective communication tools.

I can live for two months on a good compliment.

Mark Twain

When people do something especially well, give them positive feedback immediately. Relate it to the specific behavior or action involved. You will show people that you appreciate them, value them, and understand the contribution that they are making. "Prizing" them in this way will help establish rapport. The few seconds of your time involved can be extremely effective. Employees are not likely to forget your interest. Try comments such as:

> "You did an excellent job in preparing that task report last week. It was accurate and well written."
> "You were well prepared for this meeting. I really appreciate your extra effort."
> "Your speedy response pleased the customer."
> "Your consideration for them played a big part in getting this project moving."

If each of us were to confess his most secret desire, the one that inspires all his plans, all his actions, he would say, "I want to be praised."

E.M. Cioran

Many people seek recognition of their self-value and self-worth from others, even if it isn't apparent that they are doing so. Just about everyone appreciates it, whether or not they are seeking it. Prizing is an effective technique to give it to them. Be sure that your positive feedback is accurate and genuine. People receiving undeserved praise may feel better temporarily, but they will soon see through your insincerity.

There will be times when corrective feedback is required. Corrective feedback is not negative feedback if it is done correctly. Feedback to provide information for course correction is a positive action. Only if judgment enters the picture will the feedback be negative. If people need to make course corrections, you are doing them a disservice by not offering the information to them. Your silence can be construed by them as feedback that they are on course. When giving people feedback with the goal of giving them information that their behavior is not producing the desired result, give feedback on the *behavior*, not the person. Personally criticizing people is counterproductive. Comments like "You really screwed up the report" gives the person no information on the specific behavior that caused the problem.

There is a difference between effective and ineffective feedback. Here's a sample of ineffective feedback:

> "Have a chair. I'm going to give you some feedback. Several people told me that you were really rude at the meeting last week. Are you that way most of the time? If you're going to advance in this company, you'd best change and be sure that others aren't upset."

More effective feedback would look like:

"Have a chair. I have some feedback for you. Would you like to hear it?"

"Sure."

"You interrupted me several times in the meeting we had this morning. I was upset because it interrupted my presentation to the client and I was embarrassed. Can you understand how I felt?"

"Well, I can see how that might have thrown you off, but I thought you were off base and I wanted to set the record straight."

"Well, I want you to be aware of problems and correct them, but doing it in the meeting created a poor impression with the client. So, if these interruptions continue, I may attend future meetings alone. Do you have any suggestions?"

"It would help if we got together before the meeting so that we could discuss my questions in advance, and split up some of the presentation."

"Good idea. You had some good questions, but the meeting just wasn't the time to ask them. Let's plan to get together the day before the next meeting."

Table 2-1 summarizes the key differences between effective and ineffective feedback. In the first example, there were ineffective characteristics of judgment, delay, indirectness through others, not owned by the sender, no checking, referral to behavior out of the receiver's control, vague consequences, and one-sidedness. The second example avoided these problems. You'll find that others will begin to solicit your feedback when it is consistently done in a constructive and effective manner.

Every cloud engenders not a storm.

William Shakespeare

TABLE 2-1   Effective and Ineffective Feedback

Effective Feedback	Ineffective Feedback
Describes the behavior.	Uses judgmental statements.
Is immediate if possible.	Is delayed and generally comes as dumping.
Is direct from sender.	Comes indirectly through third party, i.e., so and so says you did this.
Is owned by sender. Describes the impact of receiver's behavior on sender.	Is not owned by sender.
Is checked and clarified with receiver.	Comes from sender, who is not interested in checking or clarifying.
Has specific consequences for the other person if the behavior is not changed.	Has vague consequences.
Includes asking information seeking questions to solve problem.	Is a question that is really a statement.
Is generally solicited by receiver.	Is imposed on receivers for their "own good."
Refers to behaviors that can be changed.	Refers to behaviors out of receiver's control.
Is an interactive process, and acknowledges needs of sender and receiver.	Is one sided; sender wants to win, punish, be virtuous.

Amazingly enough, nearly every negative behavior has some redeeming side to it. Recognize the positive trait of the offending behavior when offering constructive feedback. Consider an example of a person who is consistently late on project deadlines because he or she is a perfectionist who has trouble finishing because the product could always be better. Try something like, "I really appreciate the thoroughness with which you do your work. Also, we have a need to accelerate the pace to meet our customer's needs." Table 2-2 summarizes the positive qualities for several common negative traits.

Before offering corrective feedback, ask people if they'd like some feedback: "I have some feedback I could share with you. Are you open to that?" It is simply courteous to request their permission. It is very rare that people will say "no" unless you happen to have hit them at a time of great stress or emotional upset. In this case, be thankful that you asked because the feedback wouldn't have been effective at that time anyway.

Be cautious about your language to be sure that you're focusing on the behavior and not the person. Here's an example:

Attacking the person—"You're really deluding yourself. You say you're 50-percent complete with this task. You're just paying lip service to the schedule because you aren't anywhere near that complete. You may fool yourself, but you're not fooling me."

Attacking the problem behavior—"You say you're 50-percent complete with this task. When I ask about the criteria you're using to measure progress, I sense a vagueness. I hear an inconsistency in what you're saying. Can you see why I have this perception?"

TABLE 2-2   Discerning Positive Qualities Through Negative Traits

Positive Qualities	Positive Qualities Misused (Negative Traits)
Analytical	Fussiness, pettiness, over-attention to detail.
Patience	Indifferent, disinterested, permissive.
Persuasiveness	Smooth talking, pushiness, high-pressure tactics.
Frankness	Curt, blunt, brutal.
Decisiveness	Pushy, autocratic, bossy.
Confidence	Cocky, arrogant, boaster.
Flexible, Tolerant, Open Minded	Wishy-washiness, indecisiveness, spinelessness.
Thorough, Accurate	Perfectionism, impatience, rigidness.
Outgoing, Expressive, Social	Glib, melodramatic, wordy.
Resourceful, Creative, Facilitator	Over-independence, manipulating, scheming calculation.

Own your perceptions of their behavior by using terms such as, "As I see it . . . " and "I sense . . . " Do not become attached to being right. An atmosphere of right versus wrong is not conducive to effective feedback. Do you know anybody who likes to be wrong? It is far more productive to consider problems as differences of opinion, rather than in terms of who's right. Avoid win-lose situations. Nobody likes being wrong or losing.

After offering feedback, check the other person's understanding. The message you send is not always the one they receive. Ask, "What is your understanding of the feedback I'm giving to you?" Actively listening to their response will give you the information you need to correct any misunderstanding.

Receiving Feedback

I love criticism just so long as it's unqualified praise.

Noel Coward

People serve as mirrors or feedback mechanisms from which you can learn a great deal about your effectiveness. You receive many pieces of feedback every day. The key is to be open to receive them. Receiving the information that feedback offers and considering its value can be very powerful. It is only by receiving and considering the feedback that you can answer the basic questions of "How am I doing?" "Is what I'm doing working?" "What could I do that might be more effective?"

Learn to handle rejection. It's easy to handle success.

Seth Godin

A key to receiving feedback is to remember that you need not agree or disagree. It is simply information and is not to be taken personally. Do not verbally or internally attempt to justify the behavior that prompted the feedback. Be neutral (easier said than done) but achievable if you're aware of your reactions when feedback is offered. If you feel emotions coming up or an urge to justify your behavior, treat the feeling as a signal to return to neutrality. Be honest with yourself. Allow yourself time to process the information and determine its value. You may need to take some time to digest the information if you feel yourself reacting. That's a smart thing to do, not a weak thing. The only responses you need to make at the time are active listening and perception checking to be sure that you are accurately receiving the information. Once you're sure that your understanding is accurate, thank the other person for the information and go on to another topic. You can decide later how to use the feedback. Make course corrections, if appropriate. If the other person had a misunderstanding about your intent or behavior, you have gained an understanding of their point of view and can offer clarifying information.

Nonverbal Communication

What you are speaks so loudly, I can't hear what you say.

Ralph Waldo Emerson

It ain't whatcha say, it's the way howcha say it.

Louis Armstrong in the movie, "Pillow to Post"

A UCLA research study led by Dr. Albert Mehrabian found that what we say accounts for only 7 percent of what the other person believes. He found three major elements involved in the effectiveness of communication:

- What we say accounts for 7 percent of what is believed.
- The way we say it accounts for 38 percent of what is believed.
- Visual message (what the other person or audience sees) accounts for 55 percent of what is believed.

Most people worry about what they are saying, yet 93 percent of their effectiveness is not related to content. We clearly need to pay attention to the nonverbal communication that plays such a tremendously large role.

People express their emotions, desires, and attitudes more clearly through their facial expressions and body movements than they do verbally—and often more graphically. Charlie Chaplin created an unforgettable character who conveyed a wide range of emotions without saying a single word. A message delivered with enthusiasm will be more effective than a brilliantly structured message delivered in a monotone with eyes focused on the floor and no body movement. There are nonverbal gestures that have universal meaning, such as raising hands above the head to indicate surrender, saluting, shaking hands, and shrugging shoulders. Although there is debate about how precisely you can interpret body positions or movements, there is no doubt that nonverbal communication is a very strong element in any face-to-face exchange.

"You're entitled to your opinion,"
He'll tell you on the spot,
But from the way he says it,
You sorta feel you're not.

Stephen Schlitzer

You don't need training in interpreting body language to know when the verbal and nonverbal communication are in conflict. If people say they are really pleased to see you, but fail to look you in the eye, you instinctively doubt their message. If they say they are really excited about the project, but speak in a monotone with a rigid body, you wonder. Whether you are aware of it or not, you and everyone you come in contact with are constantly using nonverbal communication, both as a listener and speaker.

The face is the mirror of the mind, and eyes without speaking confess the secrets of the heart.

St. Jerome

The eyes are known as the windows of the soul and are excellent indicators of feelings. Doctors look into your eyes to check your health. Lovers stare into them to express their feelings. Moist eyes indicate the presence of strong feelings. People avoid eye contact when they are uncomfortable. Eye contact has long been considered an indicator of sincerity and honesty. Gentle, intermittent eye contact (not staring) is an indication of interest in what is being said.

The term "poker face" reflects the importance of facial expressions. It describes people who can hide their emotions by maintaining a fixed facial expression no matter what their emotion. At the opposite extreme are people of whom it is said, "You can read their face like an open book." Facial expressions are among the more sensitive indicators of a person's feelings.

Watching people's hands will also convey a lot of information. For example, tightly clenched or wringing hands are a sign of tension. Crossed arms or legs may indicate a defensive position to protect against attack or against information they are not interested in receiving. It is usually not a sign of receptiveness to communication. However, you need to evaluate the total, composite picture rather than focus on any one signal. Some people cross their arms for comfort. You need to integrate that signal with the person's rate of speech, tone, breathing, eye contact, and so forth.

Just as you are aware of nonverbal signals transmitted by others, they will be aware of those that you transmit. If your verbal and nonverbal communications are not in agreement, you'll have a credibility problem. People that are seen as open and relaxed are liked more than those who are seen as tight and closed. Managers perceived as open are more likely to get the cooperation of others. Using affirmative head nods and a warm smile when listening to others will transmit positive signals, encouraging others to be more open in their communication.

The image you project can have a major impact on your ability to communicate. The elements of image are your appearance, voice, body language, behavior, and the content of what you say. John Kennedy defeated Richard Nixon in the presidential debates of 1960 because of the more favorable image he projected, not through the content of his verbal comments. It is not often that others will be able to overcome their initial impressions of you to see some hidden talents or value. Your dress, voice, grooming, eye contact, and handshake are all elements that contribute to your image. In nearly every case, it's best to avoid extremes. For example, don't crush other people's hands in a handshake, but also don't extend a limp, lifeless hand.

Entire books are available on dressing for success. Clothing makes an important statement about you. It can provoke positive responses when properly selected, and can also be a negative factor when it isn't consistent with the message you want to deliver. Managing this aspect of your image requires knowing what image you want to present and dressing to project that image. When dealing with customers, it's important to know what they expect. For example, appropriate dress is different for a rural town council meeting than for a group of Wall Street investment brokers. Good grooming and hygiene are always important. Appropriate clothing can be offset by a layer of dandruff all over the shoulders.

The impression you make on others is based on more than the visible items of clothing, grooming, body posture, eye contact, and handshake. Your enthusiasm and cheerfulness is immediately apparent as well. If you greet everyone cheerfully and enthusiastically, you'll soon find your outlook improving or you'll drop the effort. No one can fake cheerfulness and enthusiasm for long. You'll also soon find that enthusiasm is contagious. If you can't be genuinely enthusiastic about what you're doing, it's time to reevaluate your goals and actions. It's time for a change.

Early impressions are hard to evaluate for the mind. When once wool has been dyed purple, who can restore it to its previous whiteness?

St. Jerome

Research shows that people make up their minds about other people within seven seconds of first meeting them. Much of this impression is based on nonverbal communication through our eyes, faces, bodies, and attitudes. Consciously or unconsciously, there are a multitude of signals transmitted in this brief time. Give some thought to the first impressions that you make on others. What kind of verbal and nonverbal signals do you send? What are the underlying messages sent by eyes, face, voice, and body? How much control are you exercising over these variables? They all affect your communications.

Basic Communication Tools

We've discussed the basic, important tools needed for effective interpersonal communications. Before moving on to communications via presentations, let's summarize the basic tools:

- Actively listen.
- Do not judge or evaluate the other person.
- Demonstrate empathy and understanding.
- Ask open questions.

- Check your perceptions of what others are saying.
- Check the perceptions that others have of what you said.
- Concentrate on the other person, even though your mind may want to race ahead at a speed higher than that of the spoken words.
- Give feedback effectively.
- Be generous with praise that is deserved.
- Be open to receive feedback without reacting emotionally.
- Evaluate nonverbal and verbal signals.
- Be sure that your verbal and nonverbal signals are in alignment.
- Be enthusiastic.
- Look beyond the appearance of the person for the message.
- Use language that is understandable to the other person.
- Be conscious of and control the nonverbal messages that you send.

Let's turn now to another key area of communications—presentations.

PRESENTATIONS

Oratory—the art of making deep noises from the chest sound like important messages from the brain.

H.I. Phillips

You may have to make a presentation to sell a project to a customer, to present the results of the project, to inform your project team, to teach technology, or to inform general audiences interested in your work. The same basic principles apply to every presentation, and every case is an opportunity to advance toward your goals and to grow as a person. As important as presentations are, they are the aspect most dreaded by many project managers. You aren't alone. A study by R.N. Bruskin Associates, a national research firm, found that speaking before groups was the most fearsome thing in life. Indeed, twice as many people were more afraid of speaking than *dying*. Paraphrasing Bertrand Russell, most people would rather die than speak; in fact, they do so. We'll address the issue of speaker's nerves as well as the basics of an effective presentation in this section.

Define Your Objective

The first step is to figure out why you are making the presentation, that is, to have a clear objective. Unless you have a clear objective, your message will be muddled and the opportunity will be lost. Yes, a presentation is an *opportunity*, one that you cannot afford to abuse. You must know your objective to make your point. Ask yourself: Why am I going to the presentation? What do I want to achieve? You may have only one objective. It may be to convince the customer to select you to do the project, to persuade your boss to get the computer equipment that you need, or to acknowledge your project team for their effort. Once your objective is clear, measure everything you plan to say against your objective. Your words must introduce, reinforce, or help you achieve your objective, or they don't belong in your presentation. Knowing your objective is the foundation for depth, clarity, purpose, and direction of presentation. Reduce your objective to a *single sentence* that describes precisely what you want to do. The sentence should limit your topic and define your goal.

We will use an example presentation to illustrate several of the steps involved in preparing a presentation. In this example, you are making a presentation to a local civic club on an upcoming bond election to finance improvements to the local drinking water system. The objective, reduced to one sentence, is: "To generate support for bonds for improvements to the drinking water supply system."

Define Your Audience

Did you ever try shooting at a target in the dark? You wouldn't be very successful, would you? Giving a presentation without knowing your audience has no better chance of success. Consider your audience carefully. What are their interests and attitudes? What do they want from you? What is their level of knowledge and understanding of the subject? What are their likes and dislikes? Put yourself in their shoes and ask yourself what you would want to see, hear, and feel from the presentation.

For the example presentation on the water system, the audience has only a general knowledge of the topic. You will want to avoid use of technical terms like "safe yield of the reservoir" or "optimizing coagulation." The audience will want to know why the improvements are needed and if the community can afford the costs. Their likes include enough safe water to support the community. Their dislikes include running short of water, being uncertain about its quality and health effects, and high costs. Put yourself in their shoes. This is a topic they've taken for granted. It may be unsettling to them to think that there won't always be enough safe water when they turn their tap on. They would like to be reassured in simple language that the necessary steps have been defined, are realistic, and are affordable.

If you're making a presentation in a competition to secure a project, there are several specific questions you need to address. Who is making the selection? What is their history with you and your firm? What are their personal agendas related to the project (i.e., what are each of their "hot buttons")? What do they stand to gain by the project. What is their view as to why the project is being done? What do they think of your competition? What do they perceive as the key project issues?

Get as much information as you can about your audience. Identify with them. Know what they want from the presentation.

Define Your Approach

Your approach is not the same as your objective. For example, you might have an objective of going to Los Angeles tomorrow. You can choose between approaches of walking, flying, driving, biking, and so on. The objective and approach are interdependent. If you're starting in New York, flying is the only approach that will reach your objective of being in Los Angeles tomorrow.

An approach without an objective is worthless. If all you knew was that you were leaving tomorrow, you would have no idea which approach to use. You need to know your objective—in this case, to be in Los Angeles tomorrow.

In considering your approach, know what's on the audience's minds. You may know what you want to say. Be sure you say it clearly and understandably to the people in the audience. Have an approach that makes it easy for the audience to get your message. Consider toothpaste ads aimed at parents of young children. Do they stress the scientific formula of the toothpaste? No, they focus on fighting cavities and tasting good. They're selling healthy teeth, not toothpaste. Another toothpaste ad may pursue the young adult market by showing an attractive young man and woman, to imply that their toothpaste will increase sex appeal. Be sensitive to the interests and needs of the audience in picking your approach.

Just as with your objective, boil your approach down to one sentence, after you've considered the needs and interests of your audience. This sentence should describe the single thought or message that best leads to your objective. Let's consider an example. You know that one of your customers (a large paper mill) has a new project coming up and that they are considering advertising for proposals. You have a good relationship with the customer's organization and have done good work for them. Your objective is to convince their new general manager that they should hire your firm on a sole-source basis (no competition). Your objective has been reduced to one sentence.

The next step is to learn as much as possible about the new general manager. You find by contacting several people who have worked with him in the past that he has been in the paper industry for all of his 20-year career, has worked in about every position in a paper mill and admires people who know their business, is a workaholic, is a Rotarian, recognizes that value is made up of both first- and long-term operating cost, likes ambitious people, likes to come directly to the point without chit-chat, appreciates confidence, and likes to see the people that he's going to work with.

After considering this information, you decide that this person's needs and interests include finding out how much you know about the job, some good reasons why he should give the job to you without competition, how confident you are, what the project means to you, how you've spent your career learning your business from the ground up, how ambitious you are, that you're concerned about your community as well as your job, that you can keep on the topic and come to the point, and that you are going to be directly involved in the project.

Now it's time to pick an approach. It should be based on the concept that best leads to your objective. In the above example, you could base it on the concept that your firm deserves the job because of its past good work for his company or the concept that this project will enable you to avoid layoffs at your firm. Neither would be prudent. A better concept might be to emphasize your knowledge, experience, enthusiasm, self-confidence, and demonstrated ability to do quality work for his company in a timely manner.

Let's consider the water system example presentation. You could use an approach that, unless the water system is improved, our town will be fined by the state for violating new regulations on water quality, or, if we don't expand the system, your lawns will all dry up. Neither would be productive. A better approach is: "A safe and adequate water supply is essential and affordable for our community." The best approach:

- Considers the needs, questions, and interests of your audience and how they relate to your objective.
- Develops rapport and comfort between your audience and you.
- Is the concept that best leads to your objective.
- Gives your message focus.

Developing the Presentation

You can think of your presentation being made up of three major elements—the opening, the body, and the closing.

HIX NIX STIX PIX

Headline from Variety

Did that get your attention? That's what an opening is supposed to do. The headline refers to a story about people living in the country who don't like motion pictures about rural life. Milo Frank reports that the headline sold out the issue of *Variety*. Another good example of grabbing your attention is offered by the "main titles" for television shows. The first thing you see on an adventure program is action—cars crashing, shots fired, chase scenes. The most exciting parts make up the 15 to 20 second grabbers used at the start of the show to hook you into watching. Networks pay $150,000 or more for a good main title. They do because they know they have to grab you before you turn them off. If they can hold you for a few seconds, their chances of keeping you for the whole show go up dramatically. You face the same challenge in your opening.

Your opening, like the rest of your presentation, needs to be developed to support your objective and approach. Remember, you can count on only making your one major point in the presentation. You may make several points, but they should all be in support of the one major point you want to make. After you've defined that point, you'll want to develop material that will present it in the clearest and most emphatic manner possible. Your opening must be pointed toward your objective, and it must grab your audience's attention to avoid them turning you off in their minds.

Often, people use a joke to open a presentation. It may get a laugh, but it often fails to contribute to your presentation's success because the joke has nothing to do with your objective or approach. It does little to draw the audience into your remarks.

To find an opening that will grab your audience's attention, develop one sentence answers to the questions of: What's the most unusual part of your subject? the most interesting? the most exciting? the most dramatic? the most humorous? Test these five answers against the criteria of: Does it lead to your objective? Will it relate to your listener? Does it relate to your approach? Will it excite your listener? Does it fit as the first sentence of your presentation? The sentence that best meets these tests is your best candidate for an opening.

The next step is to decide whether the opening should be a statement or a question. Usually, a question is more effective because it is more likely to get the listener's attention. Compare these two openings for a presentation from a consultant to a potential client:

- The most important thing you need in a consultant is experience.
  or,
- What's the one thing that a consultant can bring to your project to make it a success?

The second is more likely to grab the listener's attention.

Your opening needs to be brief, less than 30 seconds. It is no accident that television commercials are 30 seconds long. A lot of research has shown that 30 seconds is a typical length of time for most people to be fully attentive and open to having their attention grabbed. Milo Frank offers a good illustration of an opening used by a doctor speaking on preventive medicine:

"How would you like to die young at a very old age? Preventive medicine is the answer. Do you know that a heart attack is just your heart getting angry with you? You can keep that from happening by treating your heart well and keeping it happy. All you have to do is exercise regularly, not smoke, avoid fatty foods, and give yourself a totally relaxed day at least once a week. Do these simple things and your heart won't get angry with you. I want you to stay healthy, so don't just come to me when you're sick."

If you read the above out loud, it should take less than 30 seconds. It is an opening that grabs your attention and then gives you the core of the message.

The most effective openings combine a grabbing question along with visual impact. For example, a speaker begins his talk on how limiting beliefs affect us by kneeling to the floor and making an outline of a box with masking tape. He stands in the box and says, "What limits the success of most people? They put themselves in a box created by their beliefs about their own abilities." He steps outside of the box. "Today I'm going to talk about how to break through those limitations." That's much more likely to get your attention than if the speaker walked to a podium and said, "Thank you for inviting me. I'd like to talk about how our beliefs limit our success." Spend a lot of effort on your opening. It will pay dividends. If you don't get your audience's attention, you won't have an audience for long.

In the water example presentation, the opening might look like the following. Pour water from a pitcher into a glass. Hold the glass up, looking at it. Take a sip of water. "Can we take it for granted that there will always be enough safe water to fill our glass when we turn on the tap? Many people in this world can't. We're lucky. We've had an abundance of high quality water that makes this an attractive place to live and contributes to our healthy economy. We can continue to have enough safe water if we take the right steps. We know what the steps are. They aren't complicated, they are affordable, but they will take time. Now is the time to begin. I'll begin by showing you where we are now and where we need to go." You have caught their attention by beginning your presentation with the water-in-the-glass approach. They will be wondering what you're doing. The element of suspense will heighten their interest in your first words. In the opening, you're addressing their concerns by assuring them that there are solutions, that they are affordable, and that they are understandable. You've also conveyed a sense of urgency that now is the time to begin. In 30 seconds, you've introduced the core of your message.

I've never thought my speeches were too long; I've rather enjoyed them.

Hubert Humphrey

You've caught their attention, now you need to keep their attention with the body of the presentation. The body needs to explain, inform, persuade, inspire, and, yes, even entertain in order to keep your audience's attention. It needs to do this in the minimum amount of time consistent with your objective. You don't need a lot of words to convey an important message. The Declaration of Independence contains 300 words and the Ten Commandments contain 297 words.

He can compress the most words into the smallest idea of any man I ever met.

Abraham Lincoln

Do you know that there were two speeches given at Gettysburg, Pennsylvania, on November 19, 1863? One was by the famous orator Edward Everett. He spent days preparing and practicing his speech. The audience fell into a hush as he stepped to the podium. He spoke for over an hour. Photographers had plenty of time to record the scene. The next speech lasted less than five minutes and was only 266 words long. The photographers were still setting up when he finished, so there are no pictures. It isn't the one hour speech that is still recited today. It's the five-minute Gettysburg Address by Abraham Lincoln. It's remembered because it was heard and understood and touched the hearts of the audience. He achieved his objective in a minimum of time. You should aim to do the same with your presentation. It takes a remarkably good speaker to hold an audience's attention for more than 15 minutes. We've yet to meet anyone in an audience who has complained because a speaker finished in less time than was allotted.

No speech can be entirely bad if it is short enough.

Irvin Cobb

In preparing the body of your presentation, be sure you know your subject. Get your facts straight. Pick the aspects you want to address. You can't cover the history of the earth in five minutes. Pick a structure that best incorporates your approach and achieves your objective.

If you have an important point to make, don't try to be subtle or clever. Use a pile driver. Hit the point once. Then come back and hit it again. Then hit it a third whack—a tremendous whack!

Winston Churchill

One approach is to organize the body in three main points that explain, reinforce, and prove your main point. You might start out with your compelling point, use the second to explain it, and the third to convince the audience to act. Alternatively, you might use the first two points to set up the main point. Play with an outline until you find one that works for you.

The body should explain the point you want to make, answer the question you asked in your opening, and describe what you want done, never losing sight of the objective and approach you identified. The parts of the body of your presentation should be linked by transitions that move you from one part of the presentation to the next. They show your audience the link between the different parts of your presentation and the logical order of your presentation. The transition from the opening on preventive medicine presented earlier might look like: "Now that we've identified the basic steps in keeping your heart happy, let's look at what makes up a good exercise program."

Note that the opening to the example presentation on the water system ended with a transition to the body of the presentation: "I'll begin by showing you where we are now and where we need to go." The outline for the body of the water presentation might look like:

1. Describe the existing water system.
2. Show how the situation is changing.
3. Tell what needs to be done.
4. Show that costs are affordable.

After describing the needed actions (item 3), you transition to the cost issue by saying: "It's natural that one of your concerns is the affordability of these new facilities." You've linked the preceeding topic (new facilities) with the one to follow (costs).

A speech is like a love affair: any fool can start one, but to end it requires considerable skill.

Lord Mancroft

The last element is your closing. The closing does more than get you off stage. It can reinforce what you have said by summarizing your main points in two or three short, memorable sentences. It can invite people to act by telling the audience clearly what action you desire. It can inspire your audience by leaving them with a challenging question or a strong statement of your point. It can also set up future communication by a closing such as "I look forward to your quick approval of this proposal so that we can begin work on your project immediately." A key in developing your closing is knowing what you want from your listener. Carefully consider your listener and your objective in selecting your closing.

If you want action, let people know what is expected of them. For example, there's the story of the museum guide who was just finishing the tour, saying: "And here, ladies and gentlemen, at the close, this splendid Greek statue. Note the noble way the neck supports the head, the splendid curve of the shoulders, and, ladies and gentlemen, note the natural way in which the open hand is stretched out, as if to emphasize— "Don't forget a tip for the guide."

Maxwell Drake

If your objective is an action by your listener, you need to specifically ask for it in your closing. By asking for a specific action within a specific time frame, you're more likely to get what you want. You'll need to consider your listener in deciding just how aggressively to ask for it.

In the example water presentation, this incorporates the key elements of a good closing: "We know what needs to be done. We know that it is affordable. What is needed is your support as community leaders. I'm confident that we'll get voter support for the bond issue if we just get the story out. I'd appreciate it if you would write down one other group that should hear this presentation and drop your suggestion in the box by the door as you leave. This issue means a lot to all of us. Thanks for your help." This closing reinforces the summary of the key points—we know what to do and it is affordable. A very specific action at a specific time is requested. The value of the audience is acknowledged.

Preparing and Rehearsing

It usually takes more than three weeks to prepare a good impromptu speech.

Mark Twain

Effective presentations appear natural when you see them, but be assured that a great deal of effort goes into preparing them. Raymond Fosdick reports that Woodrow Wilson's addresses were almost invariably extemporaneous, although he prepared his outlines with care. Once, when he was asked by some undergraduates to make a speech, he inquired how long it should be. "It doesn't matter," he was told. "It matters to me," he replied. "If you want me to talk for ten minutes, I need two weeks to prepare. If you want me to speak for half an hour, I need a day. If you have no time limit, I am ready right now."

A key part of preparation is deciding on the right medium to deliver your message. Should you use slides, flip charts, video, and presentation boards, or simply deliver a verbal message? Is the message one that would be greatly enhanced by visual aids? Most are. People remember and understand more precisely what they see than what they simply hear. If you use visuals, keep them simple. If they are visuals with words or figures, never use more than five lines per slide. Convey only one message on each visual. Make sure the visuals call attention to you and your message, rather than distract your audience. The audience can get lost in studying your visual, rather than listening to you. Be sure that they fit with your objective and approach. Keep the number of visuals to a minimum. They will lose their impact if diluted by sheer numbers. We saw one speaker use 180 slides for a 20-minute talk. Most of the audience left before he finished.

Practice using your visuals so that they become a natural part of your presentation rather than appearing to be an appendage. If you're using boards or charts, mark on them during your presentation so that they become an active part of your presentation. Avoid overhead projectors. You have to stand there and feed the transparencies. You become a prop for your visual! The most common reason for using transparencies and an overhead projector is that the speaker waited until the last minute and didn't have time to prepare slides or charts. Plan ahead and use more flexible visuals.

There is real merit in writing out your speech (no, we're *not* going to recommend you read it to your audience) as part of your preparation. You've already developed an outline in one of the earlier steps. Now, write out a rough draft of your presentation, including all the facts and points that you want to make.

Avoid fancy writing. The most powerful words are the simplest. "To be or not to be, that is the question." "In the beginning was the word." "Out, out, brief candle." Nothing fancy in these quotations. A natural style is the only style.

Arthur Brisbane

Although there are about 600,000 words in the English language, the average American understands 2 to 3 percent of them. Be careful that the language level matches your audience. You don't want to speak over their heads or at an overly simplistic level. For example, a toxicologist speaking to his or her peers might say, "The flathead minnows exhibited a 100-percent mortality response." If speaking to a lay group, it would be better to say "All the fish died." Avoid jargon, regardless of the audience. Jargon can lead to the audience not understanding or seriously misunderstanding. Use interesting, simple language. Here's an example offered by Roger Ailes:

Dull (and overly complex)	*To construct an amalgam, you have to be willing to split open its component parts.*
Interesting (and direct)	*"To make an omelet, you have to be willing to break a few eggs.*

Robert Penn Warren

The quickest way to lose your audience is to use language that they don't understand.

Then, from your draft, write down key words and phrases on three-by-five cards. Memorization and recitation is not an effective method of delivering a presentation. It almost always appears unnatural and often makes the audience uneasy. You simply can't communicate with your audience if you're struggling to remember each word of your talk. Also, *never* read a presentation. You could save everybody a lot of time by mailing them a copy to read. The written and spoken word are entirely different forms of expression. Some great written literature sounds pretty stilted when read aloud. It's important to be you in front of the audience, and reading isn't the way to do it. By rehearsing from notes, you will be rehearsing delivery of the presentation naturally in your own words rather than trying to memorize the text.

Rehearse from your notes until you are thoroughly familiar and comfortable with your presentation. Try delivering your presentation in front of a mirror. It's especially important to have your opening and closing solid. Having your opening solid will help your self-confidence and ease your nervousness as you begin your presentation. Having the closing clearly in mind helps keep you on the right path.

Then, make an audio tape of your presentation. Listen to it and *time* it. You should have a presentation that is 10 to 15 percent less than your time allotted for speaking. If you can do it in even less time, great! (Remember the Gettysburg Address?) You need to have at least this time cushion because presentations almost always take longer in front of an audience than they do in rehearsal. They should because, as we'll discuss, you will be interacting with the audience and adjusting your pace to theirs in the actual presentation.

One thing an audience won't forgive is a speaker that is so rude as to arrive with a talk that is longer than the allotted time. They can't tell if the speaker didn't care enough to rehearse or knew it was too long and simply didn't care enough to edit it. By timing your tape, you'll know exactly how much time you have to cut out. Time each section. If it is too long, cut out a chunk that buys you the amount of time you need. This is much easier than changing or deleting a phrase here and there.

All the fun's in how you say a thing.

Robert Frost

Remember that only 7 percent of your communication will result from the words that you use. The rest comes from the way you say it and the way you look as you say it. Now that you're comfortable with your material, videotape your presentation. If you haven't done this before, you may be in for a bit of a shock. What you perceived as an animated speaking style may well look pretty rigid. Try watching with the sound turned off to study the style with which you are speaking. Look for distracting mannerisms, like jingling coins in your pocket, playing with your ring, or a forced smile. The most effective communicators don't change their style from private conversations to an appearance before a large group, although they may boost their energy level when delivering a presentation. Using video to rehearse is like the difference between reading about how to swing a golf club and watching a video of yourself swing a golf club. You try, you watch, and then you refine by keeping the things you do well and changing those that aren't so good. Just as in a golf lesson, you'll improve more quickly if you have a pro help you by analyzing your practices. For presentations involving the marketing of large projects, the cost of an outside "coach" may well be justified. A coach offers several advantages. They can put themselves in the position of the audience—can they understand your message? They can focus on producing the best presentation without worrying about internal politics. That is, they can tell your boss that he is mumbling, while you might be very reluctant to do so. Using a coach with communication skills and experience offers tools to use in shaping the presentation that most people don't have.

The last part of the preparation is to anticipate the questions you may be asked. First, think of the questions you would least like to be asked, the really tough ones.

There's no use avoiding them. They may well be asked. Work up responses and rehearse them. Look through your notes, putting yourself in the position of the audience. What questions would you have? List them. Have a coworker or friend do the same. With this technique, we've been able to anticipate a high percentage of the questions. You can be sure that the President of the United States and his advisors use this technique before every press conference. A key to good answers is brevity. Keep the answers to less than 30 seconds each. If the question has several parts, answer one, ask the questioner if that answered that point, and then respond to the next part. A long answer will lose the impact of your response. The questioner may well forget his question. If you drone on, others with questions will be frustrated. So, prepare *short* answers to the anticipated questions.

Delivering the Presentation

We've developed our own list of "Murphy's" Laws of Presentations:

- A question will be asked on the one topic you meant to research, but ran out of time.
- If you eat before the presentation, you will spill food or drink on your clothing.
- If you don't check your slides, they will be in backwards, upside down, or both.
- If you don't check the room, there won't be space for you, your props, and the audience.
- If you have only one projector bulb, it will burn out.
- The distance to the nearest electrical outlet will exceed the length of the electrical cords you brought for the projector.
- If you spell anything wrong, it will be your customer's name.
- All preceding speakers will run over their time allotments.
- The sound system in adjoining rooms will always be louder than the one in your room.
- Adjacent construction work will begin concurrently with your talk.
- If you don't bring a projector screen, all room walls will be dark in color.

Arrive well in advance of your presentation. Check out the room for details such as the location of light controls, whether the temperature is going to be comfortable with the room full of people (shoot for 68 degrees Farenheit), if the slide projector works, if the microphone works, and so forth. Avoid using a podium or placing anything between you and the audience. A podium may be a convenient place for notes, but acts as a barrier between you and your audience.

The greatest mistake you can make in life is to be continually fearing you will make one.

Elbert Hubbard

Now the question of last minute nervousness. As Oregon State football coach, Tommy Protho, told his team just before they went on the field for the critical game of the year, "Remember, one billion Chinese don't really give a damn how you do." This helped adjust their perspective! Another useful technique is to visualize yourself delivering a perfect presentation the night before.

Thought is action in rehearsal.

Sigmund Freud

Close your eyes, visualize walking into the room confidently and at ease. You look smart, healthy, and confident. You feel excited, comfortable, and zestful. You're introduced. From the minute you stand up, you are a powerful presence. You start your presentation perfectly, everyone sits up and is hanging on every word. You have a very pleasing, persuasive, and confident tone. There's wisdom and enthusiasm in your voice. The sincerity and humanness comes through. You are at your best. You handle all the visuals perfectly; the timing and pacing are perfect. You move effortlessly in front of the audience, totally at ease. Your gestures perfectly emphasize your main message. The audience is mesmerized. You feel a close connection and affinity with your audience; they are smiling and nodding their heads, they laugh along with you. You deliver your final message and they give you a standing ovation.

No way, you say. But we say it's foolish not to win in your own fantasies. If you do it enough times, you will convince your nervous system that you have done it for real; it doesn't know the difference and so it's old hat—no need to be nervous. So do it at least ten times the night before, then sleep on it. You will be surprised at the number of new ideas and energy you will get. Also, do it at least once just before the presentation.

Remember, the audience is on your side—they want to you to succeed. They've asked you to speak because they want to hear you. People with stage fright often tend to put their whole self-worth on the line for the 30 minutes of their presentation. Remember, there are about 720 hours in a month, 8,600 hours in a year, and several hundred thousand in a lifetime. You're life's value is not going to rest on one 30-minute period. Another technique is to consider the worst thing that could happen. You could go totally blank. The audience could laugh hysterically at your ineptitude. Your boss could fire you for embarrassing the entire company. You could be blackballed and never get another job. You could starve to death. STOP! Of course, there is no way this worst case will materialize. Although any presentation is important, it's not a life and death matter.

When you're afraid, keep your mind on what you have to do. And if you have been thoroughly prepared, you will not be afraid.

Dale Carnegie

Relax. The physiological symptoms for excitement are butterflies in stomach, increased pulse rate, agitation, and sweaty palms. They are the same for nervousness. So instead of saying "I'm nervous," why not say "I'm excited" instead? It switches the focus from fear to fun. Realize that no one in the audience knows your subject better than you do. In fact, the best cure for nervousness is your thorough preparation. If you know exactly what you're going to do when you step before the audience, your nervousness drops away. That's one of the reasons it is so vital to have your opening very clearly in mind.

When you're introduced, move briskly to the stage or front of the room. You're on stage from the minute you get out of your seat, so be prepared. Your thorough preparation should be evident by your confident manner. Project an attitude of positive anticipation. You've prepared well and you're eager to share your message. The attitude you bring to the presentation is the *most important* part of your presentation. Before you begin to speak, establish rapport with your audience by smiling and making eye contact. A study at UCLA found that 87 percent of the people researchers smiled at smiled back. Create some smiling, friendly faces in the audience. Look at your audience, not at the back wall or over their heads. Look at all of the audience so that each feels you are talking to them. You must pay attention to the audience before they'll pay attention to you. As you start your talk, look at the most friendly face in the audience. Let the eye contact register. Then, move on to another face, lock in, and read the reaction. Eye contact doesn't mean combing the crowd with your eyes like a minesweeper or going from side to side like a windshield wiper.

I do not object to people looking at their watches when I am speaking. But I strongly object when they start shaking them to make certain they are still going.

Lord Birkett

Use your eyes to absorb audience reaction and to adjust your presentation accordingly. The audience will give you continuous feedback if you look for it. Use it. As your eyes move from person to person, linger for a few seconds and talk to that person, as if you were having a conversation. Use a random pattern so that you look natural rather than programmed. If you're speaking to a group of only four or five people, you may identify in advance who you want to look at when making a specific point. For example, you may want to be looking at the accountant in the group when you are discussing your ability to meet project budgets. The next time you're in an audience, sense the difference you feel when a speaker is looking directly at you rather than elsewhere. The difference is dramatic.

We should gain more by letting ourselves be seen such as we are, than by attempting to appear what we are not.

Francois de La Rochefoucauld

For your presentation to be effective, you've got to inject yourself into the presentation. The facts you present may provide information, but it is your personal energy and emotion that provides the interpretation and much of the communication. Roger Ailes offers a simple rule: "If you have no energy, you have no audience." Move as you make your presentation, using natural movements. A static speaker produces a listless audience. As you move, the audience will be following the action, keeping in touch with you. It also helps keep you relaxed. Use your visual aids to facilitate your movement. Walk to them as you're using them, away from them toward the audience when you're finished with one. If you're giving two sides of an issue, face one way when giving one side, the other way when giving the other. Don't talk into your visuals, talk to the audience. Don't read material directly from the visuals—the audience can read. Give them information or explanations that amplify the visual, not that repeats the same information. If you are writing on a chart or board, write it first. Pause. Explain it. This sequence will keep the audience's attention far better than telling them what you're going to write before you do it.

Your movements, gestures, and posture should attract your audience to your message rather than distract them. Use a variety of facial expressions, and vary your vocal tones. A monotone is a sure sleep inducer. Let your voice express enthusiasm, variety, and sincerity. You can choreograph presentations like a dance to make your points and to tell a story. The movements are more subtle, but planning them carefully pays similar dividends.

He who laughs, lasts.

Mary Pettibone Poole

One way to make the audience comfortable is to lighten up yourself. You can take the *presentation* seriously without taking *yourself* seriously. John F. Kennedy had one of the most serious jobs in the world, but was famous for putting his audiences at ease by not taking himself too seriously:

Question: Senator Kennedy, how did you become a war hero?
Senator Kennedy: It was involuntary, they sank my boat.

Keep in mind the typical 30-second attention span. Do something different every 30 seconds, such as smiling, gesturing, moving, speaking faster or slower, speaking more loudly or more softly, asking a question, being humorous, or being dramatic. An animated presentation with variety is the goal. Remember, you are there to communicate—not to give a speech.

If there's an element of your presentation that lends itself to audience participation, do it. There's an old saying: What I hear, I forget; what I see, I may remember; what I do, I understand.

If your presentation is longer than 10 minutes, ask some simple questions at least at 5-minute intervals. Ask questions that the audience can respond to with one-word answers or by raising their hands. Always include all of the audience in your questions (e.g., How many of you have had problems filling out your tax forms? How many of you have not?) This will keep everyone feeling involved and also give you a reading on their attention. If you don't get much response, you'd better do something different because you're losing them. If you are involved in a long presentation, give the audience a break every 50 minutes.

Don't become enthralled with the sound of your own voice and deviate from your planned presentation. When you've finished your message, *quit*. Take questions.

By being prepared for questions, as we discussed, you should be looking forward to this part of your presentation. Your response to questions can often be as important as any part of your presentation. So, keep your alertness high—it isn't time to relax yet. Questions are an expression of interest. Consider every question a compliment. If you get a question you haven't anticipated and is complex or a bit difficult, try repeating the question as you heard it, or ask "Did everyone hear the question?" This will give you some time to think and will verify that you did understand the questioner's intent and that everyone heard the question. Remember, keep your answers *short*—less than 30 seconds. Ask, "Did that answer your question?", if you are concerned that your answer was too brief or if the questioner has a puzzled look on their face.

Sometimes, you may get a clearly hostile question charged with emotion. In this case, take a few steps away from the hostile questioner to give them some space. Do not move toward the person because that will intensify the hostile energy. Ask questions back until the person has vented his or her emotion and is back in control. Pose questions such as, "Would you like to expand on how you feel I was biased?" "What would I need to do to earn your trust?" "Can you give me some specifics on the reasons you disagree?" "Are you really upset over this?" This type of questioning will give the questioner a chance to vent emotions.

While the person is venting, be an observer, a Sherlock Holmes, to identify what the emotion appears to be. Is it anger, mistrust, fear, panic, confusion? Maybe it is someone playing devil's advocate to encourage you to bring out some points in your favor. It is a common trap for even experienced presenters to subconsciously feel that this is a personal attack, even though they know they should dismiss it. The presenter's level of self-esteem is the key to dealing with these situations. With high self-esteem, you will believe that the hostile questions have nothing to do with who you are as a person and does not make you a bad or unworthy person. You can accept the other person as he or she is, realizing that there is no right or wrong, only different points of view. With that awareness, it is easier to separate the information from the emotion. There may be value in the information. However, it sometimes gets discounted because the focus is on the emotion that goes along with it. It is important to use empathy in these situations: "I can understand how you feel about the high cost of this project. It must be difficult to keep to your budget and also to keep high quality, which I know is your personal goal. We want to work with you to meet your goals. What are some areas that we could reprioritize or scale back to meet the budget. Which are the most critical to you?" An empathic approach will get more cooperation than if you get defensive and play the right/wrong game. Nobody wants to be wrong, so it becomes a battle of wills. You might be right, but you might be minus a customer. Always thank people for bringing up touchy subjects, acknowledging them for the opportunity to clarify ideas that have been misunderstood.

Table 2-3 presents a checklist that is a useful guide in preparing for your presentation.

Critique Your Presentation

When it's all over, take the time to objectively review your own performance. If you know someone in the audience, ask for their feedback. If there were very few or no questions, you may have talked too long, exhausting the audience; you may never have reached the audience; or they may have gotten lost along the way and didn't know what you were talking about. Do your best to unravel why you didn't get any questions.

If the presentation was competitive to secure a project, ask for feedback from the selection panel that you addressed whether you got the project or not. If you did, find out what parts were especially effective. Keep them for future use. If you didn't, ask them where your presentation was weak and where the winners were strong. Most clients are willing to do this because they want to see effective presentations and don't want their time wasted. It also demonstrates your continued interest in them.

Compare how your actual presentation time compared with what you expected from your rehearsals. This will give you a calibration point for future presentations. Were there any visuals that seemed to work especially well? Were there any parts of the presentation that seemed to generate more audience response than others? Learning to make effective presentations is a continuous process, no matter how many you do. It's much like music:

Music is an art, not a science and no one ever becomes the greatest at an art—they could always be greater.

Arthur Rubenstein

TABLE 2-3   Presentation Checklist

Define your objective—Write a single sentence describing what you want to achieve
Define your audience:
  Level of knowledge of your subject
  Likes and dislikes
  What they want from presentation

Define your approach—Write a single sentence that describes the thought or message that best leads to your objective.

Develop your opening—30 seconds that will grab their attention. Does it:
  _____ Lead to your objective?
  _____ Relate to the audience?
  _____ Relate to your approach?
  _____ Excite your audience?
  _____ Have both verbal and visual impact?

Develop the body—Does it:
  _____ Explain the point you want to make?
  _____ Describe what you want done?
  _____ Use language appropriate to the audience?
  _____ Deliver the message in the fewest possible words?

Develop the closing—Does it:
  _____ Summarize your point in two to threee short, memorable sentences?
  _____ Ask for the action you want from the audience in terms of specific steps and schedule?

Visual aids—Are they:
  _____ Related to your point?
  _____ Helping you make your point?
  _____ Readable from all parts of the room?
  _____ Compatible in number with the time available?

Timing:
  _____ Have you timed your presentation?
  _____ Is it at least 10 percent shorter than the allotted time?

Dreaded questions:
  _____ Have you listed and answered the questions you'd least like to be asked?

General questions:
  _____ Have you listed and answered all the questions that you can anticipate?

Room layout—Is the room and audience size compatible?

Other distractions—Have you checked out the lighting, temperature controls, and possible adjacent sources of noise?

Projectors/visual equipment—If you're using a projector, do you have a spare bulb, extension cords, and screen?

Slide test—Have you projected all of your slides in advance to be sure they are right-side up?

Choreography—Have you planned how you will move to and from the visuals?

TABLE 2-3   Presentation Checklist *(continued)*

Audience participation—Is there an element that you could use for audience participation?

Question the audience—Have you identified questions that you could ask the audience during your presentation?

Handouts—Are there any written materials you want to have available to distribute at the end of the presentation?

Post presentation—After the presentation, analyze:
What worked?
What didn't work?
Presentation time vs. rehearsal time

References

1. Rogers, C.R. 1980. *A Way of Being.* Boston: Houghton Mifflin Company.
2. Rogers, C.R., and F.J. Roethlisberger. 1990. Barriers and gateways to communication. *People, Managing Your Most Important Asset.* Harvard Business Review. Boston. p. 19.
3. Hulnick, R., and M. Hulnick. 1990. Unpublished course materials, University of Santa Monica, California.
4. Hunsaker, P.L., and A.J. Allessandra. 1980. *The Art of Managing People.* New York: Simon and Schuster.
5. Qubein, N.R. 1983. *Communicate Like A Pro.* New York: Berkley Publishing Group.
6. Hoff, R. 1988. *I Can See You Naked—A Fearless Guide to Making Great Presentations.* Kansas City, Missouri: Andrews and McMeel.
7. Ailes, R. 1988. *You Are The Message.* New York: Bantam Doubleday Dell Publishing Group.
8. Frank, M.O. 1986. *How To Get Your Point Across in 30 Seconds or Less.* New York: Pocket Books, Simon and Schuster.
9. Graham, R.J. 1989. *Project Management as if People Mattered.* Bala Cynwyd, Pennsylvania: Primavera Press.

3

Motivation and Leadership

MOTIVATION

We know nothing about motivation. All we can do is write books about it.

Peter Drucker

Much has been written about motivation. Although there is no simple formula or universal approach for achieving motivation, we know that motivation plays a critical role in achieving project success. Everyone is motivated to do something. The challenge is to find out what it is and where the individual goals overlap with your project goals. Project success occurs when the project is led rather than managed and when the leader and team are motivated.

An Internal Process

There is not one of you who has not in his knapsack the field marshal's baton; it is up to you to bring it out.

Louis XVIII, speaking to the
Saint-Cyr Cadets, 1819

You can't unilaterally motivate others. External or extrinsic motivators, such as rewards or punishment, produce only short-term results. As soon as the reward becomes expected or the punishment disappears, so does the motivation. Some of the most motivated people in the world, such as Mother Teresa and her volunteers, are obviously not motivated by external rewards. True motivation comes internally:

Motivation = A Heartfelt Mission Based on Values

Strongly motivated people follow their inclinations toward a mission that has strong personal meaning to them. There is a difference between results based on missions and ones based just on goals. If you have no deep sense of personal mission, you can still check off a list of goals accomplished and feel good or satisfied about it. However, unless they are related to a deeper mission, your desire will eventually decline and your results suffer. Motivation reaches a peak when your personal goals and mission align with the specific demands of your work and the overall objectives of your organization. You like what you're doing. You are committed. You feel yourself growing. You see concrete results. You feel a strong sense of inner purpose. This is what motivation feels like—and it can come only from within.

You don't motivate others; instead, you encourage and allow people's motivations to be directed toward the project goals. You can promote a project environment and provide leadership that will encourage motivation. We devote the rest of this chapter to ways of doing this.

People's Needs Differ

Practice without belief is a forlorn existence. Managers who have no beliefs but only understand methodology and quantification are modern-day eunuchs. They can never engender competence or confidence.

Max DePree

Nearly 50 years ago, Abraham Maslow proposed a model for motivation. We have learned a lot in the subsequent years. Yet Maslow's model underscores an important fact—you must know each person to understand their needs and to know the environment that will stimulate them.

To a man with an empty stomach food is God.

Mahatma Gandhi

Maslow's model states that people are motivated to fill their needs in an ordered sequence or hierarchy. For example, people quickly forget about food if deprived of oxygen. At the first level, they seek to fill their physical needs, such as food, shelter, water, and air. Hungry people will focus their activities on finding food, and they will give little energy to other needs. Once the first level of needs are continually satisfied, more food or shelter will have no motivational value. They then focus on higher level needs. Satisfied needs no longer motivate people.

Second level needs revolve around safety, security, stability, and freedom from threats to one's physical well being. The third level involves social needs, such as friends, a sense of belonging, and family. The fourth level involves esteem-related factors, such as self-respect, recognition, attention, and appreciation by others. The fifth level addresses self-actualization, self-esteem needs, the needs for self-fulfillment, and to be able to grow and learn.

We know, as did Maslow, that these steps do not always have to be fulfilled in sequence. A starving artist offers the perfect example of someone sacrificing lower level needs in pursuit of the higher levels of creativity. However, in general, Maslow's model is a useful reminder that you must evaluate the needs of your team members on an individual basis. Some may be seeking status or recognition, while others are primarily seeking fulfillment of social needs, such as a feeling of belonging.

Recognize that people are motivated by differing considerations and will be motivated by their own individual desires. Get to know each member of your team and understand their needs. You then have a chance to give them project assignments in which they can become motivated by satisfying their needs. A blanket assumption that all will be motivated by one thing, such as salary increases or performance bonuses, will assuredly lead to failure. People are motivated more by feelings and sensitivities than they are by facts and logic. They quit high-paying jobs because they don't feel appreciated, don't feel challenged, or don't like the work environment. Michael Le Boeuf boils down the keys to maintaining a work environment that encourages long-term motivation for people to this: "They belong to an organization that cares about them, challenges them, believes in them and wants the best for them, not just as employees, but as total human beings."

Management Styles Affect Motivation

Many years ago, Douglas McGregor identified two managerial styles and how they affect motivation: Theory X and Theory Y.

Theory X assumes that people dislike work, avoid responsibility and work, look to management to make decisions, and require coercion and direction to produce. The Theory X approach does not provide an environment conducive to motivating team members!

I am your King. You are a Frenchman. There is the enemy. Charge!

Henry IV, Battle of Ivry, 1590

Just because you're king doesn't mean that you have to be an autocratic, Theory X-type manager. King Arthur gave his knights plenty of chances to participate as they sat around his round table. The shape of the table was no accident: he wanted to encourage participation.

Theory Y assumes that people take joy in doing good work, seek responsibility, want to improve, want to be self-directed, want to be active, and can be creative, and that managers can rely upon workers self-direction.

I neither ask nor desire to know anything of your plans. Take the responsibility and act, and call on me for assistance.

Abraham Lincoln to General Grant
on his appointment to
command the Union Armies, 1864

The assumptions that you make about your team tend to determine their behavior. If you believe that they are lazy and do not want responsibility, you are likely going to watch them closely and tell them how to do their work. Over time, they will appear to behave as you expect. They will not be willing to take responsibility because they know you are likely to change their decisions. If you believe they are self-directing, you would tend to leave them alone and their behavior would be quite different. Theory X and Theory Y are simplistic representations of the extremes of a scale of management styles. The theories illustrate the effects that your beliefs can have on your team's motivation. You can create a self-fulfilling prophecy where your expectations of your team lead to a behavior that meets your expectation.

Theory Z describes still another concept. According to Theory Z, motivation in an affluent society such as in the United States occurs because workers are driven by higher needs such as:

• Self-esteem
• Being proud of their organization
• Fulfilling one's potential
• Fully participating

With Theory Z the manager aims to fill these needs by offering autonomy and responsibility at work. Theory Z is aimed at realizing the motivational power unleashed when individual and team goals come into alignment. It simply suggests that involved workers are the key to success.

Much work on management styles has been done since McGregor's famous work. The important factor is to recognize that your style of leadership will have an effect on the motivation of your team. Few of you are likely to fall into the extreme of McGregor's Theory X style. However, if you recognize this as your style, it's time for a change! We'll soon discuss leadership skills that will enhance team motivation.

Satisfaction vs. Dissatisfaction Factors

With work which you despise, which bores you, and which the world does not need—this life is hell.

William Edward Burghardt DuBois

Frederick Herzberg made extensive studies to determine what motivated workers. He found that distinctly *separate* factors lead to job satisfaction and job dissatisfaction.

In other words, the opposite of job satisfaction is not job dissatisfaction, because the factors involved are different. Rather, the opposite of job satisfaction is simply no job satisfaction. He separated this into "motivator" and "hygiene" factors. The hygiene factors that can cause dissatisfaction (but not satisfaction) include company policies and administration, supervision, relationship with supervisor, interpersonal relationships, working conditions, salary, status, personal life, and security.

The motivators that cause job satisfaction in decreasing order of frequency cited were achievement, recognition, work itself, responsibility, advancement, and growth. Herzberg listed some principles or actions involved in applying these motivators, as shown in Table 3-1. Herzberg's motivators correspond to the higher levels of Maslow's hierarchy.

Money is what helps people get to sleep at night, not what gets them up in the morning. Success and accomplishment are what really motivate people.

Linda Honold

How Important Is Job Satisfaction?

To sense the potential of posterity in one's daily business life, one has to feel deeply about some aspect of one's work. One has to be committed to do or create something that will accomplish some good in the world.

David Finn

TABLE 3-1   **Applying Herzberg's Motivators (From Reference 3)**

Action	Motivators Involved
Removing some controls while retaining accountability.	Responsibility and personal achievement.
Increasing the accountability of individuals for own work.	Responsibility and recognition.
Giving a person a complete natural unit of work (module, division, area, task, and so on).	Responsibility, achievement, and recognition.
Granting additional authority to employees in their activity; job freedom.	Responsibility, achievement, and recognition.
Making periodic reports directly available to the workers themselves, rather than to supervisors.	Internal recognition.
Introducing new and more difficult tasks not previously handled.	Growth and learning advancement.
Assigning individuals specific or specialized tasks, enabling them to become experts.	Responsibility, growth, and advancement

One study reported by Schmidt and Posner found that 50 percent of U.S. managers surveyed said their careers gave them the most satisfaction in life. Family was a close second at 40 percent, with outside interests providing the remaining 10 percent. This information shows that there is a tremendous source of psychological energy available if you can tap this desire to be fulfilled. There is a tremendous inclination for people to be motivated by their work under the right conditions. A vice president of General Motors said it well:

> . . . all people are at their best when they are the essential members of an organization that challenges the human spirit, that inspires personal growth and development, that gets things done and that symbolizes and stands for only the highest standards of ethical and moral conduct. That is what quality of work life is all about.

If people gain only financial rewards and meet lower level needs, they will never contribute more than their minimum effort. It is by employing their heads and hearts that they become motivated. Max DePree, chairman of Herman Miller, Inc., provides an insightful summary:

> What is it most of us really want from work? We would like to find the most effective, most productive, most rewarding way of working together. We would like to know that our work process uses all of the appropriate and pertinent resources: human, physical, financial. We would like a work process and relationship that meets our personal needs for belonging, for contributing, for meaningful work, for the opportunity to make a commitment, for the opportunity to grow and be at least reasonably in control of our own destinies. Finally, we'd like someone to say "Thank You."

How many times have you taken other people's efforts for granted or been in too much of a hurry to thank them? Make a conscious effort for a day to thank people. Watch the reactions.

It is through the art of leadership that you can meet the needs for members of your project team and organization, and provide an environment where they are likely to be motivated.

LEADERSHIP

Managers vs. Leaders

People cannot be managed. Inventories can be managed, but people must be led.

H. Ross Perot

Management is doing things right; leadership is doing the right things.

Peter Drucker

Managers honor stability, control through systems and procedures, maintenance of the status quo, and often act with little passion or involvement. Leaders bring out the best in people, don't give up on people, have a high degree of self-confidence, and guide people to new places. Leaders have looked within themselves to know who they are and what they want their team to be. They passionately pursue their visions.

Leadership appears to be the art of getting others to want to do something that you are convinced should be done.

Vance Packard
The Pyramid Climbers

The distinction between managing and leading is the distinction between getting others to do and getting others to *want* to do. Managers have used extrinsic rewards and pressures such as promotions, bonuses, and titles to get others to do things for many decades. If the positive doesn't work, managers resort to demotions, firings, or putting people on probation. Leaders get people on their team to want to do by establishing a clearly stated, inspiring vision arrived at jointly with the team, by empowering the team by expanding their authority, and by continually communicating his or her heartfelt belief in the project mission. Leaders do not change people's values, but rather uncover what people will commit themselves to. They articulate what people have felt but haven't said and establish a vision that focuses their attention.

Kouzes and Posner surveyed 2,615 top-level managers to find out what they thought were characteristics of a superior leader. Here are the top 15 characteristics and the percentage of managers surveyed who listed the characteristics.

Characteristic	Percent
Honest	83
Competent	67
Forward Looking	62
Inspiring	58
Intelligent	43
Fair-Minded	40
Broad-Minded	37
Straightforward	34
Imaginative	34
Dependable	33
Supportive	32
Courageous	27
Caring	26
Cooperative	25
Mature	23

Effective leaders then:

- Are honest
- Are competent
- Have a long-term future orientation
- Have a clear, strongly held sense of direction
- Communicate their vision to others, appealing to a common purpose
- Have energy and enthusiasm springing from a strong belief in purpose
- Have a positive, can-do attitude
- Don't control others, rather trust them and enable them to act
- Believe in people's potential
- Are in close touch with those they lead
- Care about others
- Get everyone involved
- Respect the aspirations of others
- Behave as they want others to behave—they walk like they talk
- Are active, not passive
- Experiment and take risks
- Consider mistakes as learning opportunities
- Welcome change
- Are good listeners, are open to the ideas of others, and encourage contrary opinions
- Achieve a team result beyond the sum of individual contributions
- Recognize and celebrate the contributions of others

A leader is not an administrator who loves to run others, but someone who carries water for his people so they can get on with their jobs.

Robert Townsend

Leaders Establish a Vision

The very essence of leadership is [that] you have to have a vision. It's got to be a vision you articulate clearly and forcefully on every occasion. You can't blow an uncertain trumpet.

Theodore Hesburgh

Have you ever tried to put together a jigsaw puzzle without having the picture on the box to work from? We have participated in group demonstrations where there are three teams putting puzzles together under the following conditions: 1) the first team has the picture of the completed puzzle on the box; 2) the second team has no picture on the box; and 3) the third team has a puzzle that doesn't match the picture on the box. None of the teams are told anything about their puzzles or pictures. No surprise—the team with the picture on the box always completes the puzzle in the least amount of time.

Every project begins as a vision in someone's mind. A leader paints a picture of this vision so that the project team clearly sees what the completed puzzle will look like and where each of their pieces will fit. A leader holds the vision intact and keeps communicating it to the team with spirit and enthusiasm that conveys the leader's heartfelt *belief* in the project mission. Without effective communication of the vision, the project values will be buried by trivial reports and memos.

If you don't know where you are going, you will probably end up somewhere else.

Dr. Laurence J. Peter

Vision for a project usually begins with a somewhat vague feeling or desire to accomplish a challenging goal. By definition, establishing a vision involves looking into the future. As the vision becomes clearer, you get a sense of what the final picture will look like and perhaps even some ideas of how to get there. Having the picture of where you are going is the key. You can't get from here to there if you don't know where "there" is. Effective leaders project themselves ahead in time to clearly see the final picture.

The best way to create the future is to invent it.

Alan Kay
Apple Computer

John Kennedy had a clear picture of an American standing on the moon by 1969. When his vision was clearly defined and communicated, a massive effort by highly motivated teams set about to accomplish one of man's most spectacular achievements. The characteristics that caused this vision to be so motivating are common to all motivating visions. Kennedy's mission was a challenge. It set a goal that was beyond normal capabilities. Although much of the technology didn't yet exist, it was not so far beyond our capabilities as to be hopeless. It offered a chance to be tested. It was a chance to change the way things were. The vision was broad enough that it allowed the teams to use their own ingenuity in solving problems. It provided a chance to develop and learn. It was unique—it would put America on the moon first. It would give the team great pride. It looked to the future. It provided a chance to *make a difference*. The vision gave everyone on the team a chance to do something, rather than merely having something to do. Project contractors reported amazing changes in attitudes and intensity when their staff began working on the project. The changes transcended day-to-day concerns to uplift the entire team. These are the essentials of a vision with high motivation potential. It attracts people to it by the force of its appeal.

Vision is the art of seeing things invisible.

Jonathan Swift

There are few missions as glamorous as going to the moon, but there are many examples of visions of individuals leading to great success. Steve Jobs and Stephen Wozniak had a vision of a personal computer. They sold Jobs's VW bus and Wozniak's calculators to start Apple Computer Company. Their vision and leadership led to products that all of us are familiar with. Ted Turner had a vision of a 24-hour news network. Many thought him crazy, giving him little chance of success. The strength of his vision and leadership created the Cable News Network, now taken for granted.

Managers fight fires . . . leaders light fires.

Perry Pascarella and Mark Frohman

As well as defining the mission, you must communicate the vision and light as many fires as you can. If others do not see hope for achieving their own hopes and desires, they won't be motivated. So, to communicate the vision, you appeal to a common purpose, communicate clearly, and sincerely believe in the vision. People are very adept at seeing through a fraudulent message. Leaders are committed to a vision that they are passionate about and, as a result, produce the results of leadership. They make sure that their vision comes alive and stays alive by continually communicating their message and keeping in close contact with their team.

Leaders are very sensitive to the collective and individual needs of the team so that they develop a sense of what the team members value and of a common purpose. They identify the individuals they want to enlist in their vision. They get out of their office to spend time with them, to ask what they want from the project. They listen.

The key to success is simple. Make people dream.

Gerard de Nerval

Leaders communicate the vision in language that paints a clear picture of it. Martin Luther King's *"I Have a Dream"* speech is a classic example. Consider the following excerpts:

> I have a dream that one day on the red hills of Georgia the sons of former slaves and the sons of former slave owners will be able to sit down together at the table of brotherhood.
> I have a dream that one day the state of Alabama, whose governor's lips are presently dripping with the words of interposition and nullification, will be transformed into a situation where little black boys and black girls will be able to join hands with little white boys and white girls and walk together as sisters and brothers.

King's language painted vivid pictures. It was positive and hopeful. If you've ever seen a film of his speech, you know that he felt deeply about his beliefs. The language made it clear that he was convinced that his vision was achievable. He appealed to the common purpose of his audience. He used examples that everyone could understand and relate to. It was more than the impassioned manner in which it was delivered that motivated people. It was the clarity and appeal of the message.

Your adrenaline has to run, whatever business you are in, if you don't feel exhilarated by achieving your objectives and excelling in what you're doing, then you will never do very much well.

Malcolm Forbes

Leaders deliver their messages with enthusiasm and a positive can-do attitude. People will not follow someone with a negative attitude. They express warmth and friendship, not an aggressive attitude, in their communications. Their heart-felt commitment to the vision permeates their communications. They are more compelled to serve their vision than their own ego.

Leaders Challenge Limiting Beliefs

Consistently we observe that the weakest muscle in the body is the one between the ears. Self-imposed limitations and beliefs hold most people back.

Ralph DuBois

Until 1954, no one had run a mile in less than four minutes. There were a number of medical papers that said it was physiologically impossible—man's lungs, circulation, and muscles simply weren't designed to run a mile in less than four minutes. Such learned discussion should surely have discouraged Roger Bannister, a young medical student. All of history and medical literature said it couldn't be done. Yet, in 1954, he ran a mile in 3 minutes and 59.4 seconds. He not only broke the world's record, but broke a barrier imposed by a limiting belief. Once the barrier was broken, many other runners began to run less-than-four-minute miles. The barrier wasn't physical, it was mental.

Leaders find opportunities for people to explore new ground. They question the status quo. They establish a mission that is not bound by preconceived limitations. By doing so, people will go beyond their fears of failure to reach new heights.

A man flattened by an opponent can get up again. A man flattened by conformity stays down for good.

Thomas Watson, Jr.

You can attack the limiting beliefs on your projects by looking at the ways the work is being done. Ask, why are we doing it this way? Is it really contributing to our being the best we can be? Is it stimulating creativity? If not, change it.

Ask employees what's wrong, not what's right.

An Wang

Look for the cultural practices in your organization that restrict thinking and change them. Don't accept that it's the way it's always been done. Leaders change the rules to fit their own vision. Ask the project team what needs to be changed. Ask them why they are doing what they do. Devote at least 25 percent of every team meeting to brainstorming innovative approaches. Give everyone a chance to participate in shattering a limiting belief by coming up with a new approach, procedure, or process. Chapter 9 offers some specific tools you can use to involve team members in brainstorming and evaluating ideas for changes.

The Wright brothers flew right through the smoke screen of impossibility.

Charles F. Kettering

Leaders Take Risks

I failed my way to success.

Thomas Edison

The Chinese ideogram for crisis consists of the characters for "danger" and "opportunity." Leaders take risks and encourage others to do so—not foolhardy, blind risks, but educated risks that build self-confidence. John Kennedy knew there were risks in setting a national goal of walking on the moon by 1969; he also knew the chances of success were reasonable. He knew that the gains from a sense of mission and accomplishment would result in an increase in national esteem and confidence that justified the risk.

Success does not breed success. It breeds failure. It is failure which breeds success.

M. Maidique

Risk is inherent in any project that is going to make a difference. A leader makes it safe for team members to take risks (i.e., making it safe to fail and learn from mistakes). If the above quote strikes you as illogical, consider Babe Ruth. You no doubt know him for his success—714 home runs—but he also struck out 1,330 times. Babe lived the advice he offered others, "Never let the fear of striking out get in your way."

If you miss seven balls out of ten, you're batting three hundred and that's good enough for the Hall of Fame. You can't score if you keep the bat on your shoulder.

Walter B. Wriston

Leaders persist even when risks lead to failures. Alexander Graham Bell failed to get any financial backing for his invention (the telephone) for years. Chester Carlson's dry copying process was rejected by many large companies before he eventually obtained financing and succeeded as Xerox. Abraham Lincoln failed in business in 1831. He was defeated for the Illinois state legislature in 1833. His sweetheart died in 1835, and he had a nervous breakdown in 1836. He was defeated for Congress in 1843 and, after being elected in 1846, lost his Congressional seat in 1848. He was defeated for the Senate in 1855, lost out for the vice presidency in 1856 and was defeated again for the Senate in 1858. Today, there is a memorial in Washington to this great President.

Remember your past mistakes just long enough to profit by them.

Dan McKinnon

A leader accepts honest, reasoned failures as the price for innovation. Their team members are not afraid to keep looking for a better way to do things. The leader keeps the focus on what can be learned from the failure. If your project team is spending their time and energy fearing your reactions, their imagination and creativity will be inhibited.

You can encourage risk by honoring your risk takers. Management consultant Tom Peters, suggests an innovators Hall of Fame in your company where you display risk-oriented project material such as pictures, plaques, prototypes, and so forth, for all to see. Reward the good tries as well as the successes. Successful leaders and team members are driven by a strong desire for achievement, not by a fear of failure. There is a big difference. They expect to occasionally get off course. The only way to avoid getting off course is to stand still—and you can't progress toward your mission if you're standing still.

Meet your failure nobly, and it will not differ from success.

Ralph Waldo Emerson

You can help your team deal with their fears of failure. It's helpful to keep in mind the following way of looking at "Fear" as an acronym:
F alse
E xpectations
A ppearing
R eal

I've had a lot of problems in my day, most of which never happened.

Mark Twain

A University of Michigan study found that 60 percent of our fears are totally unwarranted; 20 percent have already occurred and so are beyond our control; 10 percent are so petty as to be insignificant; and only 5 percent of the remaining are justifiable fears. Half of that 5 percent are things we can't do anything about, which leaves 2 percent that we can solve if we just start acting or planning instead of fearing.

Bad times have a scientific value . . .We learn geology the morning after the earthquake.

Ralph Waldo Emerson

As a leader, the way you address risks that don't work out is important. First, you need to understand the background of the action. Was the reason for what was done arrived at by a reasonable thought process? If so, you should offer words of positive reinforcement. The words you use and the way you deliver them will have a big impact on people's willingness to take risks in the future. Be sure to separate the performance from the performer. Don't say, "You really didn't get the results we wanted." Praise the performer at every opportunity. Even if test results weren't what was hoped for, you can say, "The tests were run exactly as we discussed. I appreciate the effort you put into the project and the precise way you did the work. You're a great example for the team. Let's check out what happened together." Focus on learnings, not the problems, to create winners.

Success is on the far side of failure.

T.J. Watson
Founder, IBM Corporation

It's important to analyze the failures as well as successes in order to learn from them. Focus on the shortcomings in the system (see Chapter 9 discussion on systems concepts) that contributed to the failures, not on individuals. Work with your team to identify what could be done better next time. Record the notes and make them available to everyone. Review them before you start the next project.

Our greatest glory is not in never falling but in rising every time we fall.

Confucius

Leaders Are Honest, Establishing Trust

There is one cardinal principle which always must be remembered: one must never make a show of false emotions to one's men. The ordinary soldier has a surprisingly good nose for what is true and what is false.

Field Marshall Erwin Rommel

Honesty topped the survey of characteristics people want to see in their leaders. Before following you, your team wants to know that you are worthy of their trust. How do you demonstrate honesty? *Do what you say you are going to do.* This can be as simple as showing up on time for a meeting you promised to attend. It can also be keeping people on the payroll in tough times when you promised a no-layoff policy.

Example is not the main thing in influencing others. It is the only thing.

Albert Schweitzer

People trust others when they follow through on commitments and behave in a straightforward manner. Consistency between words and actions are vital. If you as a leader say one thing and do another, your honesty rating plummets. The next time you make a promise, no one will believe it. You also gain trust by making your values, ethics, and standards clear to your team. As a leader, you are respected when you take a stand on your values. You are also trusted when you are open, as opposed to keeping your cards close to your chest. You are more likely to be trusted if you trust others.

If you want a man to be for you, never let him feel dependent on you. Make him feel you are in some way dependent on him.

General George C. Marshall

You need to demonstrate your commitment and trustworthiness before asking for the same from someone else. If you want someone to do something for you, do something for them. Such reciprocity leads to predictability and trust. You can also recognize the value of interdependence by realizing that people working with you will accomplish far more than you can alone, physically and mentally. You will find that you can depend on your team because they can depend on you. You demonstrate trust when you delegate effectively and encourage people to partici-pate. Leaders have a genuinely positive view of others and are willing to share responsibility. You are willing to consider their view-points and to use their abilities to the fullest. You create a trusting relationship when you show others that you depend on them and that your success depends on them. People on your team exercise a lot of responsibility because they respect you and each other.

In fact, I've come to the conclusion I never did know anything about it.

Thomas Edison, on electricity

Be open and honest about your own mistakes and limitations. People don't believe or trust you if you claim to be infallible. Admit your mistakes, make appropriate corrections, and keep driving toward the vision.

Trust men and they will be true to you; treat them greatly and they will show themselves great.

Ralph Waldo Emerson

For project success, trust is an essential ingredient. Psychologist Carl Rogers has observed that trusting another's competence, judgment, or concern results in a greater willingness to be open with that person. Feeling trusted also makes it easier to be open. Projects succeed when team members and the project manager are open with each other, exchanging ideas in an environment where they feel safe to speak freely.

It's important that people know what you stand for. It's equally important that they know what you won't stand for.

Mary Waldrop

To establish trust, you must know your values and live them. Your team needs to know what you stand for. Defining your values (such as teamwork, quality, innovation) lets your team know what they are supposed to do. Your team will give much more weight to the values that you use as opposed to those that you only speak about. If you're preaching teamwork, abolishing all reserved parking spaces will underscore that you mean it. Your values are demonstrated by how you spend your time. If you preach the merits of high-quality work, your credibility will go up if you are actively involved in quality management training and in performing quality reviews. Also, your values are reflected in the questions you ask. If you ask only about deadlines and budgets, your team will conclude you're not really concerned about quality. If you were, they would expect you to ask about how your customer is feeling about the project. They also watch what you reward. If you say you encourage innovation and risks, promote and reward those that take reasoned risks, even if they fail occasionally. If a failure is punished, your credibility is destroyed. If you don't set out clear expectations, your staff will be second guessing you. They will not be able to meet your expectations unless they know them. They will become frustrated and demotivated.

A thousand words will not leave so deep an impression as one deed.

Henrik Ibsen

Leaders are Competent

Do your homework. You can't lead without knowing what you're talking about.

George Bush

Before you'll follow someone, you have to believe that they know what they're doing. It does not necessarily mean that the leader has to have outstanding technical abilities in the basic technology of the business. You don't need to know how cows convert grass to milk to manage a dairy farm. Lee Iaccoca probably has a very limited grasp of all the technologies involved in manufacturing an automobile, yet he is widely accepted as an outstanding leader of Chrysler Motor Company. General Groves certainly did not understand the theories of atomic physics, yet he provided the project leadership that resulted in producing the atomic bomb that ended World War II. The key expectation that must be met is that you can get things done for your project team that must get done. Technical competence may be necessary, but, in itself, it is not enough. You need to demonstrate the ability to effectively plan, encourage, coach, and inspire the team in order to be seen as competent. Over time, your track record becomes a demonstration of competence.

Be constantly alert for skills that you lack, and eagerly develop or acquire them. Project managers often find that their interpersonal skills can be improved with substantial benefits to their teams and to themselves. The most successful managers pay attention to the human side of their projects. Yet most do not have training in the needed people skills, and it is very rare to be born with them. Participating in training courses to improve your interpersonal skills demonstrates to others that you are serious about improving the team atmosphere.

Another area that can frequently be strengthened is project planning. The larger the projects you lead, the more important your planning abilities become. Without them, you will appear (and likely are) incompetent to achieve your project team's objectives.

The only things worth learning are the things that you learn after you know it all.

Harry Truman

The most effective leaders have a never-ending impulse to learn and grow, ever increasing their areas of competence. As Maslow said, you have capacities clamoring to be used which cease their clamor only when they are used sufficiently.

Leaders Align Individual and Project Missions

You must capture and keep the heart of the original and supremely able man before his brain can do its best.

Andrew Carnegie

Leaders appeal to the intrinsic motivation of others. A tremendous amount of energy is released when the project goals and the goals of the individuals on the team are in alignment. You need to know what the individuals want and make project assignments that match their needs and goals. Randolph and Posner found that the unwillingness or inability to understand the perspective and needs of others is the primary reason that project managers fail. Several researchers have found that the ability to understand and be sensitive to the goals of others is a key to project success. Alignment occurs when an individual feels that his or her contribution to the project is contributing to his or her personal goals. The more opportunities that his or her project assignments allow for such alignment, the better the chances for success.

Some guys play with their heads, and sure you need to be smart to be number one in anything you try. But most important, you've got to play with your heart . . . a guy with a lot of head and a lot of heart, he'll never come off the field second.

Vince Lombardi

Athletic teams offer a simple example of the concept. Both the team and individual have the goal of being the best. The individual wants the team to win because it provides the vehicle for achieving his own goal. A superior coach comes to know each player well enough that their team assignments best match their skills *and* desires. Getting team and individual missions in alignment produces amazing results by getting both the minds and hearts into action. As a result, team results can exceed the sum of the individual abilities. The Boston Celtics basketball team won 16 titles without ever having the league's individual leading scorer.

Charles Garfield has suggested that the opposite of the Peter Principle (people rise to the level of their incompetence) occurs when individual and team goals align. He refers to it as the Paul Principle: a position of considerable competence from which all things seem possible. When you experience this alignment, you are committed and you feel yourself growing and learning. You are in a place to make your mission happen, and you see results from your work. You feel strong because you are contributing your best. Focus on the Paul Principle.

A leader gets to know his or her people and their goals. The active listening skill we described in Chapter 2 is an essential skill. It must be combined with spending the time with your team in order to listen to them. Ask each of them about their goals and needs. Determine what part of the project most appeals to each of them. If one has a goal of professional recognition, give that person the assignment of preparing any published papers on the project. If one's goal is to get experience in all phases of design and construction, give that person the construction assignment that they have not yet had. Their energy for the construction assignment will be much greater than if they were given an assignment involving more of the design work that they have already done.

Never try to teach a pig to sing; it wastes your time and it annoys the pig.

Paul Dickson

Trying to force-fit someone to a project assignment that is not in alignment with their goals is unproductive. If there is no fit on the project, recognize it and look for another assignment for the individual.

Leaders Get People Involved

. . . we are most at leisure when we are most intensely involved.

Marshall McLuhan

The most productive project teams are ones where everyone is important, not just the project manager. When team members feel that their project manager is strengthening them and encouraging collaboration, their level of satisfaction and commitment increases. Leaders are not a crutch for their team members to lean on, but are catalysts for growth and development. Commitment to a project comes when people are involved. Unless they are involved, they simply won't be committed. The process can begin by getting the entire team involved in developing a statement of the project mission. During the project, use teams to address specific project issues. Get input from the "doers." They know their tasks better than anyone else. If you're trying to do it yourself, you're not going to get the best answer or the best project performance. Involvement is fostered when you encourage the team to share information, listen to each other, exchange resources, and respond to each other's needs. Enable individuals to make decisions. Create opportunities for interaction between the individuals on the team. Chapter 9 discusses practical applications of team involvement.

No great idea ever entered the mind through an open mouth.

James Kouzes and Barry Posner

Seek many inputs on project issues. Practice the active listening skills described in Chapter 2. You can't incorporate people's ideas and views into the project unless you know what they are. When they are incorporated, team interest and commitment rises. Share or give credit for ideas.

Keep *everyone* on the team informed about the project goals, status, problems, and successes, as well as where each person fits into the overall picture. The more they know, the better off you and the project will be. With information, they will take on more responsibility and have more resources to achieve successful results.

Encourage your team members to attend professional conferences, take training courses, and visit peers with other companies to keep their level of professional involvement high. While cost is always a consideration for training, there are many trainings that have a minimal cost for a day. Maybe you can negotiate with the team member to make up 50 percent of the time if there is a clear individual benefit. You may be surprised about how little it takes to motivate people when you both have an investment in their growth. If they ask to attend a training session, find a way to encourage them. Training helps keep teams motivated by showing them that they are growing, learning skills to take on new challenges, and making more contributions. Such activities will also increase their technical knowledge. Teams that are cut off from outside communication are cut off from valuable information and can become unimaginative.

By encouraging involvement in outside professional activities, you'll be keeping your team open to new ideas and information. Encourage involvement in activities that build interpersonal skills, such as Dale Carnegie courses and leadership positions in civic and volunteer organizations.

My philosophy is that you can't do anything yourself. Your people have to do it.

Beth Pritchard

Leaders recognize that project goals are only fully realized through the efforts of all, not just a few or by their own effort. They know that the involvement of all in an environment of openness, fairness, respect, and honesty is a key to project success.

Leaders Are Active, Not Passive

Be willing to make decisions. That's the most important quality in a good leader. Don't fall victim to what I call the "ready-aim-aim-aim-aim syndrome." You must be willing to fire.

T. Boone Pickens

Effective leaders gather the best information available, make a decision, and move on. They recognize that triumph is "umph" added to "try." You don't sit around waiting for opportunities or for perfect, complete information. You take the initiative and do whatever is necessary and proper to get the job done. You consider decisions that others find stressful as an opportunity to develop. You expect to influence the results. People will not follow those who will not take decisive action.

The important thing is poise. How a man handles a situation is a much more important thing than the situation itself. Poise in all things and at all times. So few men have it.

Lord Northcliffe

Effective leaders are also active rather than passive in regard to ownership of their own behavior. You know that your behavior is a result of your own conscious choice, rather than a result of an uncontrollable reaction to external circumstances. Your behavior is a function of your decisions and not outside conditions. It's not what happens to you that determines how you feel, but rather what your response is.

How you react to the issue is the issue.

You actively own and choose your reactions. For example, you may have, at some time, said something like, "I am upset because Bill is late with his progress report again." Such a passive, reactive approach puts you in the role of a victim, not in charge of your own destiny. Such an approach puts other people or circumstances in charge of your situation. Recognize that your upset is a mental, emotional response that isn't related to the physical world. It is a response that you can control by your own choice. You recognize that it's not the event that causes you upset, but your reaction to the event that causes your response. Therefore, "I am upset" is the only important part of the sentence. Ask yourself, "Why am I upset?" Have you noticed that the same event has happened on other days and you didn't get upset? What other factors were involved: you just had an awkward call from a customer; you shouted at your kids this morning and feel bad about it; the person you're mad at reminds you in some way of another person you have had a bad experience with in the past. We tend to react out of habit based on conclusions drawn from our experience. We sometimes project these experiences onto other people and expect the worst or overreact. You have taken a big first step if you become aware that it is an internal process of yours that is causing the upset, *not* the external event. The next step is to choose to react differently based on the individual situations.

No one can make you feel inferior without your consent.

Eleanor Roosevelt

By choosing to respond in a constructive way to an event, an effective leader breaks the connection between outside events and his or her responses.

Leaders Encourage Contrary Opinion

Where all think alike, no one thinks very much.

Walter Lippman

A team of similar-thinking people can develop blind spots. It's a phenomenon called "groupthink." It's generally believed that President Kennedy failed to get enough input, especially contrary opinion, before deciding to go ahead with the Bay of Pigs invasion of Cuba, which failed. He changed his approach before dealing with the Cuban missile crisis and sought a diversity of opinion.

We owe almost all of our knowledge not to those who have agreed but to those who have differed.

Charles Caleb Colton

Differing opinions and their free expression is good for the health of the project. A leader recognizes the diversity of people's talents and opinions as a resource, not a liability. They encourage the expression of contrary opinions and actively listen to them. They encourage differences and discourage indifference. You can encourage people by asking them if they see anything wrong with an approach or if they have another idea on how to approach it. Urge them to be the devil's advocate. Maintain a safe environment for them to respond and give careful consideration to their opinions.

Your people aren't giving you their best . . . if they are afraid to speak up.

Kenneth Labich

Leaders Establish Doable Goals

Beware the big play: the 80-yard drive is better than the 80-yard pass.

Francis Tarkenton

Nothing is particularly hard if you divide it into small jobs.

Henry Ford

Leaders get people to tackle big challenges by breaking them down into smaller, doable tasks. It's the same approach that behavior modification programs like Alcoholics Anonymous, Weight Watchers, and Smoke Enders use. Alcoholics Anonymous doesn't insist that you become totally sober immediately. They ask you to stay sober one hour at a time, then one day at a time, as you work toward your long-term goal of continuous sobriety.

Psychologists have long recognized the value of a series of small wins. Leaders recognize small wins. With every small win, more forces are set in motion that encourage steps toward another small win. Your team's confidence level will rise as their desire to feel successful is reinforced. The small wins deter opposition. After all, it is hard to argue with success. Resistance to later proposals drops. Your team will see that you are asking them to do things that they know they are capable of doing.

Project milestones are an example. A complex project is broken down into a series of small tasks, each with its own description and schedule, or milestone. As one task builds upon another, the overall project moves to completion. Trying to accomplish too much at once or presenting an overwhelming task in one bite doesn't work. It's like learning to downhill ski. You don't start on the advanced slopes. You start on the beginner area. Once that is mastered, you move on to the "green" slopes. Eventually, you reach the "double diamond" slopes. If you started there, you quickly would have given up—and probably ended up in the hospital!

Pick battles big enough to matter, small enough to win.

Jonathan Kozol

The leader is sensitive to the capabilities of the team and the limits of their resources. A key to project success is to divide the project into small, doable tasks and give your team constant feedback as each task progresses. They must know how they are doing before they tackle the next task. You wouldn't expect a basketball team to be able to adjust their strategy if they didn't know the score until the game was over.

Leaders Recognize Performers

Pass the pride down. People like to create when they can earn recognition for their ideas.

James L. Hayes

Old praise dies, unless you feed it.

English Proverb

Much like the quarterback who bought Rolex watches for his linemen after winning the Super Bowl, you can demonstrate how much you value your team members' contributions. If the true performers receive no recognition, they may well lose interest or move on to another project or organization. It is important for you to define what is expected of people and provide them feedback. As we discuss in Chapter 9, formal performance appraisals are a terribly ineffective way to recognize performance. Leaders recognize the performers in a variety of ways, including extrinsic rewards such as bonuses, raises, or promotions. They also address the intrinsic needs for recognition. Recognition, especially when spontaneously offered immediately after an outstanding performance, is a powerful tool.

Do not neglect gratitude. Say thank you. Better still, say it in writing. A simple note of thanks is money in the bank and you will be remembered.

Princess Jackson Smith

Try the simple phrase "thank you." To be recognized and appreciated is a basic need. Leaders motivate people to major accomplishments through the simple acts of recognition and praise.

Rewards need to be based on the needs of the person you're recognizing. What's rewarding for one person may be seen as a punishment by someone else. To a hungry person, food would be a reward. To someone already full (or on a strict diet), being forced to eat would not be viewed as a reward. Tune in to the needs of the individual involved before deciding on the form of recognition.

Positive recognition is a powerful motivator if it is sincere and directly related to a specific accomplishment. Undeserved praise is demeaning and damaging to the person receiving it. People resent receiving praise from someone who has no idea of what they've done. Also, be specific. Don't just say "good job" or "keep up the good work." Rather say something like, "Sarah, this is another outstanding report. You really amaze me. I want you to know how much I appreciate working with someone as effective and personable as you."

The first thing you have to learn in leadership is that there's no room for ego.

Jon Levy

The famous football coach Paul Bear Bryant said he learned how to hold a team together by saying just three things:

If anything goes bad, I did it.
If anything goes semi-good, then we did it.
If anything goes real good, then you did it.

There are times when recognition of the entire team is appropriate and valuable. Word that your proposal for a new project has been accepted or the completion of a major milestone justifies a project team celebration. At times like this, leaders become cheerleaders. One bank placed a large bell in the middle of the office. Whenever one of the staff closed a loan, they got to ring the bell. Mervyn's department store executives send out note cards that have a heading, "I heard something good about you" at the top to pass on compliments. The Tennant Company uses "That-A-Way" notes—brightly colored three-inch by five-inch cards that employees give to each other to recognize superior performance on a task. The CEO of Versatec personally cooked dinner at the home of the top-performing employees. Sam Walton, chairman of Wal-Mart Stores and one of the richest men in the country, told his employees that he would put on a hula skirt and dance on Wall Street if they met their profit goals. They did, and he did. Such public recognition can do much to strengthen team commitment and enthusiasm.

There are two things people want more than sex and money . . . recognition and praise.

Mary Kay Ash

If you have any doubt about this quote, look at the results that Mary Kay Ash has achieved. She has built the Mary Kay cosmetics idea into a company that grosses hundreds of millions of dollars per year. Her system provides super-recognition for super performance. In the mid 1980s, her top 50 sales representatives averaged over $100,000 per year, but the real key is the annual Awards Night. On that night, 8,000 people who sell Mary Kay products are showered with applause, gifts, and praise. The most striking gift is a Cadillac in the company color—pink. A large number of sales and recruitment achievers are crowned, complete with flowers and scepters, while receiving standing ovations and musical fanfares. Mary Kay attributes her success to making the people on her team feel important. You can do the same for the people on your team without resorting to gifts of Cadillacs. Public praise for good work, congratulatory letters, favorable publicity, and the types of public celebrations described earlier are all effective tools.

Leaders Make It Fun

People are never so trivial as when they take themselves very seriously.

Oscar Wilde

If your team isn't enjoying what they're doing, your project is suffering. Doing boring, routine tasks in a solemn atmosphere does not produce the best results. The leader gets rid of routines that don't serve the project mission. The passion with which the leader works toward his or her mission excites others. It fuels the fun of testing skills and being creative. A leader makes the project enjoyable by offering team members challenges and opportunities to do different things and explore new areas, and by demonstrating a sense of humor. Although there were several other factors, the differences in the sense of humor between John Kennedy and Richard Nixon accounted for some of the differences in their reputations as leaders. One project manager energized his team by taking them to Disneyland for the day after they had finished one phase of the project. What could be a Disneyland equivalent for your team?

Projects are fun when the team members are growing, personally and professionally. People enjoy their work when they have a clear vision of where they're headed and the knowledge that they will help shape their own future. The leader provides the environment for the growth and shares in the excitement.

This chapter has described a philosophy with hints and examples to give you the sense of experience and environment that you, as a leader, could create. Here is a summary:

- When you assemble your team, find out about their needs. Plan your strategy to satisfy these needs (include your own).
- Hold a project kickoff meeting. Include everyone. Act as facilitator to develop a

mission statement—do not dictate. Set the example up front. Discover their needs by asking questions and listening. Describe your expectations for the project, the team, and individuals. Develop a consensus of expectations. Then ask for team commitment and support. Develop a project ideal scene for expectations in positive, present tense (see Chapter 5).

- Hold project team meetings regularly. Give notice of them, plan for them, delegate parts of the agenda to team members, and get the doers involved. Use the Total Quality Management tools (Chapter 9) to solicit information and ideas from all members. Such an approach takes the pressure off you for infallible decision making and gets the team involved. Encourage members to challenge beliefs, take risks, and learn from mistakes. Maximize the chances for small wins to build confidence and trust between you and team members.
- Keep your commitments/agreements. Do what you say, or renegotiate ahead of time.
- Share all information in meetings or circulate information. Surprises will eat away at the trust level.
- Encourage "storming," whether it be brainstorming or conflict. It is healthy and avoids groupthink.
- Sit down with the team and individuals to establish doable goals. Get their input. Time spent in this process will be more than saved in reduced rework time or time spent heading down blind alleys.
- Reward the team and individuals (and yourself) when you reach milestones—pizza for lunch, a team party, whatever it takes. Practice saying "Thank you."
- Make it fun—it'll make your life easier, as well as your team's.

References

1. Kouzes, J.M., and B.Z. Posner. 1987. *The Leadership Challenge.* San Francisco: Josey-Bass, Inc.
2. Rosenau, M.D., Jr. 1984. *Project Management for Engineers.* New York: Van Nostrand Reinhold.
3. Herzberg, F. 1990. One more time: How do you motivate employees? *People: Managing Your Most Important Asset,* Harvard Business Review, Boston.
4. Katz, R.L. 1955. Skills of an effective administrator. *Harvard Business Review* February 1955.
5. Garfield, C. 1986. *Peak Performers.* The *New Heroes of American Business.* New York: William Morrow and Company.
6. DePree, M. 1989. *Leadership is an Art.* New York: Bantam Doubleday Dell Publishing Group, Inc..
7. Slevin, D.P, and J.K. Pinto. 1988. Leadership, motivation and the project manager. *Project Management Handbook,* ed., D.I. Cleland and W.R. King. New York: Van Nostrand Reinhold.
8. Hill, R.E., and T.L. Somers. 1988. Project teams and the human group. *Project*

Management Handbook, ed., D.I. Cleland and W.R. King. New York: Van Nostrand Reinhold.

9. Miller, T.E. 1988. Teamwork—Key to managing change. *Project Management Handbook,* ed., D.I. Cleland and W.R. King. New York: Van Nostrand Reinhold.

10. Thamhain, H.J. 1988. Team building in project management. *Project Management Handbook,* ed., D.I. Cleland and W.R. King. New York: Van Nostrand Reinhold.

11. Wilemon, D.L., and B.N. Baker. 1988. Some major research findings regarding the human element in project management. *Project Management Handbook,* ed., D.I. Cleland and W.R. King. New York: Van Nostrand Reinhold.

12. Maslow, A.H. 1943. A theory of human motive acquisition. *Psychological Review,* p. 370.

13. Schmidt, W.H., and B.Z. Posner. 1983. *Managerial Values In Perspective.* New York: American Management Association, p. 23.

14. Covey, S.R. 1989. *The Seven Habits of Highly Effective People.* New York: Simon and Schuster.

15. Waitley, D. 1983. *Seeds of Greatness.* New York: Simon and Schuster.

16. Pascarella, P., and M.A. Frohman. 1989. *The Purpose-Driven Organization.* San Francisco: Josey Bass, Inc. Publishers.

17. Hersey, P. 1984. *The Situational Leader.* New York: Warner Books.

18. Ouchi, W.G. 1981. *Theory Z—How American Business Can Meet the Japanese Challenge.* Reading, Massachusetts: Addison-Wesley Publishing Company.

4

Negotiating

YOU NEGOTIATE WITH PEOPLE

Everybody's negotiable.

Muhammad Ali

Each of us negotiates something every day. We may negotiate a multi-million dollar contract or where to go to dinner. Negotiation is communication intended to reach agreement between two parties that have some interests in common and some that are opposed. Even if the parties are large companies or countries, negotiations always occur *between people.* It is individuals that determine group or company goals. In fact, no matter how large or complex the other side's organization appears, your chance of success will be better if you look at the problem from the point of view of the one person that you are dealing with.

The ability to deal with people is as purchasable a commodity as sugar and coffee. And I pay more for that ability than for any other under the sun.

John D. Rockefeller

You deal with people who have emotions, values, and motivations—just as you do. Each of you see the issues from your own perspective. It is easy for your perceptions to be inaccurate. You may not understand what the other party means from the words they use. They may not see the meanings you intend in your words. Misunderstandings can trigger emotions and reactions that sabotage the negotiations. By maintaining a person-centered approach that deals with others sensitively as human beings with human reactions, you can effectively negotiate. The basic communication skills discussed in Chapter 2 are key tools to use in negotiations.

79

NEGOTIATIONS SHAPE PROJECT QUALITY

Once selected to do the project, people are usually eager to get on with the work. Project contract negotiations are often viewed as a necessary step, but one that should be as quick and painless as possible. After all, the buyer of your services has selected you to do the work. Both of you are eager to get on with the project. However, both should recognize that negotiations of project scope, price, terms, schedule, and other factors are key steps in determining ultimate project quality and client satisfaction. In the process of negotiations, the client and you develop a mutual detailed understanding of the project. It is an excellent opportunity—and certainly the right time—to be sure that you truly understand the client's needs and desires.

Negotiations, in many cases, establish working relationships for the project. You can use the negotiation process to develop mutual respect and trust and to learn the most effective way to communicate. You can learn how to best work together. The project budget, schedule, price, and key personnel assignments are developed during negotiations. By jointly developing these key project elements, both parties will have a greater sense of commitment.

Negotiations establish the framework for the project. They must be treated with the same level of respect, effort, preparation, enthusiasm, and knowledge as the marketing of the project and the actual project work. This chapter first addresses some of the concepts underlying successful negotiations and then describes some of the specific issues and steps involved.

DEFEAT THE PROBLEM—NOT EACH OTHER

If you're going to skin a cat, don't keep it as a housecat.

Marvin Levin

The goal of negotiation is not to kill the other party, but rather to achieve an agreement. A negotiation that results in an overwhelming victory for one side is likely to be short lived and certainly not conducive to a long-term relationship. A danger in the win-lose approach is that even if you win the first round, the other party may be looking for and get a win in the next negotiation on the same project. If you can see yourselves working together to attack the issue and not each other, the chances for success are greatly increased. The chances diminish if you treat the issue and the people as one. People should be dealt with as human beings and the issue on its merits. We'll talk about dealing with the issue in the next section. First, let's look at dealing with the people aspects.

An adversarial frame of mind in negotiations leads to many difficulties. Anything that one person says about the issue is taken as a personal affront by the other.

Each side becomes defensive and reactive. The issues can be lost in the resulting personal conflict. Typically, when entering negotiations, you see only the merits of your position and the faults of the other.

The ability to see and feel the issues as though you were the other side is a valuable skill. This doesn't mean you have to agree with their point of view. However, you do need to go beyond just understanding that they have a different view—it is experiencing the emotions surrounding the issue as they feel them and living their point of view. There is an old Indian saying that you don't criticize another until you have walked in his moccasins. It doesn't say until you've studied his moccasins. You need to know what it feels like to be in them.

Do everything you can to understand the other person's point of view—it takes the wind out of the sails of a blustering manipulator. If you can demonstrate that you're trying to understand their needs, eventually they will run down and you will be able to assert your point of view.

Robert Decker

By living the other's point of view, you'll gain another perspective on the issue. It can be helpful to have someone role-play the other side's position. Even though you are not in total agreement, you may change some of your views, reducing the areas of disagreement, improving the chances of success, and gaining rapport. It also is productive to discuss each other's perceptions of the issues because it can provide insights essential to agreement.

Language is important as you discuss perceptions to avoid placing blame. For example, saying that "your analysis of this information is inaccurate and your conclusions are wrong" is going to be taken as a personal attack. Rather, "we must be looking at this information differently. I feel that . . . " will convey your perception without appearing to be a personal attack. The basic communication skills discussed in Chapter 2 will all come into play during negotiations.

People are much more likely to accept an agreement if they have had a hand in developing the solution. It will pay dividends to get the other side involved early. The chances for success will get stronger as each side has input on parts of a solution. Give the other side an opportunity to discuss a draft agreement before you get to the negotiation table. Agreement becomes easier if both sides feel some ownership.

It is a fact of life that emotions are a factor in negotiations. Chapter 2 discusses communication skills that are effective in dealing with emotions, such as perception checking, reflecting feelings, and active listening. A key in negotiations is to see that emotions are expressed. Unexpressed emotions can be a major burden. If you sense that the other side is feeling angry, check your perception by asking, "I sense that you are angry. Am I correct?" If the other side explodes in an emotional outburst, let them vent their emotions freely. Silence can be very effective in eliciting information in such a situation. Listen quietly without responding or becoming defensive. Do not interrupt them.

The greatest remedy for anger is delay.

Seneca

The toughest part of this approach is that you are likely to feel attacked while they are venting. Always remember that you are allowing the other person to relieve their own stress and frustration. It is not a personal attack on you. Separate the emotion from the information. The emotion is theirs. The information is for you to evaluate.

Likewise, if you are feeling angry, discuss it by saying something like, "I am feeling that our views are not being listened to and I am upset." Be careful to speak from your perspective about a problem's impact on you, rather than what the other side said or did. Saying "I feel let down" is more productive than saying "You broke your word." If you attack them with a statement they consider inaccurate, they will be upset or ignore you. A statement strictly about your own feelings can't be viewed as inaccurate and can open a useful dialogue. Directly describing and talking about emotions can diffuse an explosive situation. By airing the emotions, you can return to addressing the problem.

The best way of answering a bad argument is to let it go on.

Sydney Smith

As the speaker's emotions begin to subside, they are more likely to keep talking and give you information if you are silent. The silence is likely to make them feel uncomfortable if they do not offer more than their emotions. If appropriate, ask the speaker to continue until they are finished. Give them encouragement to speak out without inflaming the situation. Be attentive. Separate the emotion from the information in their message. If you're unclear about the information, you might say, "I understand how you could feel that way, but I'd like more clarity about the information. Would you please tell me more?" No matter how rough the rhetoric gets, it will be counterproductive to take it personally and react emotionally.

When you say something, make sure you have said it. The chances of your having said it are only fair.

E.B. White

Don't take communication for granted in negotiations. In the best of circumstances, people often misunderstand another's intent. In negotiations, people may be feeling hostile and suspicious, further clouding their perceptions. You may be so busy figuring out how to put forward your own position that you fail to listen. Applying the active listening skills discussed in Chapter 2 is important. Check your perceptions by asking the other side, "My understanding is that you feel the delivery date is unrealistic. Is that correct?" Improve communication by letting the other side know that you are really hearing them. They will feel heard and understood. Unless you demonstrate that you understand what they are saying, they may believe that you have not heard them. They will discredit your point of view because they will believe that you haven't yet understood theirs. If you accurately paraphrase their case before discussing it, you'll have a better chance of constructively talking about the merits of the problem. They will know there is no misunderstanding of their view. Also, listen carefully for what is *not* being said in negotiations. Broad general statements are a clue that perhaps specific issues of substance are being avoided. Ask open-ended questions to elicit information.

Take whatever opportunity you have to get to know the person you're negotiating with in advance of the negotiations. Visit them in their office prior to negotiations. Their books and personal items in their office will give you useful clues for conversation topics to get acquainted and establish rapport. If you can deal with someone you know, the negotiations will be easier. You'll have a better understanding of their motivations, a basis of trust, and some established basis of communication. When you get to the negotiating table, try some symbolism to indicate you're working side-by-side on solving the problem. Sit on the same side of the table with the draft contract in front of you so that you jointly face the common task.

How well the other person identifies with you can be a major factor in the negotiations. Act in a professional and responsible manner, and speak to their needs, hopes, and aspirations to develop empathy. Treat them on the human level, demonstrating a sincere interest in solving the problem at hand.

DEALING WITH THE PROBLEM

How many a dispute could have been deflated into a single paragraph if the disputants had dared to define their terms.

Aristotle

The key to dealing with the problem is to determine and understand the interests of each party. Sounds simple, doesn't it? However, negotiations often break down because people focus on the positions involved rather than the underlying interests. The proverbial quarrel over an orange illustrates the point. Two men wanted one orange. They finally agreed to divide the orange in half. One ate the fruit from his half and threw away the peel. The other used the peel from his half to bake a cake and threw the fruit away. If they had defined their interests rather than their positions, both could have been more satisfied. Too often, we assume that if another person's position is opposed to ours, then their interest must also be opposed. As illustrated by the orange, this isn't necessarily true. Of course, the trick is to figure out the underlying interest.

One method is to put yourself in their shoes and ask why they have taken a certain position or why they are not agreeing to your position. Look for interests related to the basic human needs of security, economic well being, recognition, a sense of belonging, and control, because they will be the strongest. Don't assume that financial interests are all that are involved. For example, focusing solely on salary when negotiating with an employee will not be productive if his real interest is the recognition that could come from being a vice-president. Certainly, negotiations won't progress if one side feels its basic needs are being threatened by the other. Negotiations with staff about a new quality management program may not progress until they are assured that no one will lose their employment as a result of ideas from the program.

The most direct way to explore interests is to talk about them so that each side understands the importance and legitimacy of the other's interests. The other side will be more likely to appreciate your interests if you show that you appreciate theirs. People feel that those who understand them are people who are worth listening to. Apply perception checking skills by asking such questions as, "As I understand it, your interest is using the orange peel to bake a cake and you have no interest in the fruit. Is that correct?" A question like that would have made both parties in the earlier example happier with the outcome. There is little merit in being fixed in your position—I want the orange. There is merit in being firm on your interests—I want the fruit, not the peel. If two people each state their interests and push for them, they will often stimulate each other's creativity in coming up with mutually beneficial solutions. You can push hard for your interests while still treating the other person as a human being and being open to their input. In fact, you can't expect the other person to listen to your interests unless you demonstrate that you have heard and are considering theirs.

Once the interests are defined, it's time to develop specific options to address them. It is important to have a range of options identified prior to beginning negotiations. For example, before beginning compensation negotiations with a potential employee, it would be wise to brainstorm various options so that you will be prepared and have thought through all the implications before the negotiation. For example, options could include a company car, education assistance for a graduate degree, stock options, salary versus bonuses, club memberships, extra vacation, added insurance coverage, and so forth. In negotiating a contract, options for tradeoffs between scope, schedule, and cost should be thought out in advance. Getting your team together to brainstorm options in advance of negotiations will give you a broader base to draw upon.

In evaluating options, it is advantageous to find those that provide mutual gains through either different or shared interests. More for you does not necessarily mean less for me. Remember the quarrel over the orange? Each could have had the part of the orange they wanted if they had only identified what they wanted. Identifying shared interests can also lead to workable options. That new employee may be concerned about security. You are concerned about making a big investment in the new employee because he or she might walk out the door next year. Both have a shared interest in some form of assurance that the relationship will not be temporary. An option of a three-year employment contract could address this shared interest.

An effective method is to offer options that are acceptable to you to see which options the other side prefers. For example, you could offer to produce a report in six months at a cost of $75,000, or in nine months at a cost of $70,000; you could offer to evaluate three alternatives instead of five; you could ask the customer what he or she prefers. You can continue to refine the options until you find no more joint gains.

PLANNING AND PREPARING FOR NEGOTIATIONS

Before everything else, getting ready is the secret of success.

Henry Ford

Successful negotiations are 70 percent preparation, 20 percent implementation, and 10 percent acting.

Robert Olson

Learning how to negotiate by trial and error can be a very expensive education. An ill-prepared, inexperienced negotiator facing a prepared, experienced negotiator is headed for serious trouble. There is much that can be done prior to negotiation to level the playing field.

Deciding on the Right Negotiator—or Team?

The basic first step is to decide who is going to negotiate. Will it be an individual or team? If an individual, who specifically?

There are tradeoffs when selecting a team versus an individual. The team approach has several advantages. You will have a greater pool of expertise to draw upon, as well as a greater variety of perspectives. It may be difficult to find one person with all of the expertise needed to make knowledgeable decisions or present reasonable arguments for negotiating a large, complex project. Also, since each person on the team will be viewing each issue from their own unique perspective, more creative solutions may result. You will also have a greater number of ears to listen with, making it more unlikely that a comment will be missed. With multiple sets of notes, you'll have more complete and accurate minutes of the negotiations. There is also less of a chance that your side will make an error—be it in judgement or a simple arithmetic error. A team can provide the intangible benefits of moral support. In some circumstances, there may be a psychological advantage in having numbers of negotiators in your favor. The group approach allows more members of the project team to be involved. This accomplishes two things: 1) more of the team gets a chance to get to know the client before the project starts; and 2) their sense of ownership of the project approach increases because they are involved in developing it.

There are also disadvantages to the team approach. The amount of coordination increases as the number of participants increases, exponentially according to some who recommend three team members as a maximum. It is unlikely that all will have the same understanding of the project. You will have to devote time and energy to briefing the team. Unless responsibilities are clearly defined, each team member may think that someone else on the team will take responsibility for the effort; the overall effort will be weak if no one takes the lead. It is also possible that disagreements among your team members could surface during the negotiating sessions, allowing the other side to take advantage of the lack of unity. The other side may play the interests of one of your team members against the rest of your team. The members have probably not worked together as a team before and may not behave in a manner so that their individual skills meld together effectively. The supportive nature of a group can lead to a feeling of invulnerability, leading to undue risk taking in negotiations. Fear of looking weak to team members may lead to avoiding concessions that should have been considered.

There is obviously no universal answer to the team versus individual approach. You'll need to consider the complexity of the project, the approach of the other side (are you dealing with a multidiscipline team or an individual), the time available to prepare, the costs of negotiation versus the potential profit on the project, and the skills of the individuals available. What are the characteristics of an effective negotiator? They include:

- **Patience**—Most people want to end the tension of the uncertainty associated with negotiations as quickly as possible. A patient negotiator has an advantage. We are not suggesting that you unnecessarily prolong the negotiations, but that you have the patience to listen and to carefully think out options without rushing into agreement.

With time and patience the mulberry leaf becomes a silk gown.

Chinese Proverb

- **Well prepared, knowledgeable**—You should know the project thoroughly, from the technical, legal, and fiscal perspective. You should know the client's needs, goals, and characteristics well. You need to recognize the limits of your knowledge and be open to calling in expert help when appropriate.

- **Good business sense**—You need to be able to relate issues to the bottom line of project schedule and profit. If the other side raises a question or makes a proposal, you should be able to quickly analyze its effect on the project costs and profit. We have seen cases where a technically-oriented project manager eager to reach agreement has given away any hope of profit—and not even realized it.
- **Tolerates conflict and uncertainty**—Often times, many issues are discussed but are best not resolved until other related issues are resolved. For example, it isn't prudent to define price until the project schedule is finalized. There may be several parallel issues, all of which remain unresolved for a substantial part of the negotiations. Nothing should be considered final until all items are negotiated. The negotiator needs to expect this and be comfortable with the resulting uncertainty. You also need to have the ability to remain calm and reasoned during periods when the other side may become emotional.
- **Willing to take reasonable risks**—Every negotiation involves some risk taking. You may be torn between setting an easy goal where the risk is small or establishing a lucrative goal with a greater risk. Any one person's willingness to take risk will depend on a number of factors. How much work do they already have? How secure are they in their abilities to get other work if the negotiations fail? Yale University conducted an interesting study of high achievers. They found that high achievers are not big gamblers seeking the thrill of hitting a long shot, nor are they happy sticking with a virtually sure thing. They prefer to take risks with about a 50-percent chance of success. A good negotiator can strike a balance and select an appropriate risk level. A reasonable risk taker views the challenge of negotiation as an opportunity rather than a threat or problem.
- **Able to relate on a personal level**—The negotiator should be willing and able to get involved with the individuals on the other side on a personal basis. As we've already discussed, the ability to relate can make or break a negotiation. You need to understand them in order to develop solutions that satisfy their needs.
- **Good listener**—Chapter 2 and early parts of this chapter have stressed the importance of effective listening. You must be able to listen to the other side to understand their goals and needs in order to negotiate an effective agreement.
- **High self-esteem**—Having a strong need to be liked by others is not a good trait for a negotiator. By having high self-esteem, you are not going to give away the store in order to be liked by the other side. You are less likely to be influenced by the other side and more likely to tolerate ambiguity.
- **Sense of humor**—The ability to see humor in the negotiating atmosphere can go a long way toward preserving your own mental health. Humor can also be an effective tool in negotiations by relieving the tension of the group. In one case, a negotiator stood to leave, picked up his briefcase, which was unlocked, and the entire contents (seemed to be a multitude of small items) spilled across the table

and floor. He quickly said that, as a last resort, he'd decided to spill his secrets. The room erupted in laughter and, because he had a good sense of humor, so did he. The tension was relieved and negotiations resumed.

- **Integrity**—An unethical, unscrupulous negotiator will soon lose the respect of the other side and destroy the effectiveness of negotiations. You must have a strong sense of integrity.

I'd rather be the man who bought the Brooklyn Bridge than the man who sold it.

Will Rogers

- **Stamina**—Negotiations can become lengthy. People who tire easily can make big mistakes as negotiations drag on. You need to be in reasonably good physical shape to be effective over a long period of time. Negotiating with a hangover or serious jet lag is a mistake. Arriving a day before negotiations to prepare and rest is a prudent investment if long distance travel is involved. We have been involved in several negotiations assisting an attorney who must have one of the largest bladders and smallest stomachs on earth. He seems to never need a break. His ability to stick with an issue until it's resolved is a major advantage. The ideal negotiator can maintain a high level of mental and emotional activity over a period of several hours by being in good condition.
- **Persistence**—Most people are not persistent enough in negotiations. If the other side rejects a proposal, they accept rejection and walk away. History abounds with examples of the value of persistence.

Results? Why man, I have gotten a lot of results. I know 50,000 things that won't work.

Thomas A. Edison

Thomas Edison had 14,000 failures before he found a successful filament for the electric light bulb. Dr. Seuss's first children's book was rejected by 23 publishers. The 24th publisher sold six million copies. During his first three years in the automobile business, Henry Ford went bankrupt twice. President Carter's persistence is given much of the credit for the Camp David agreement between Egypt and Israel. For 13 straight days, he kept Anwar Sadat and Menachem Begin locked up at Camp David, refusing to accept anything less than an agreement between the two longstanding enemies. He got it. Although some people claim that anyone would be willing, after 13 days locked up in Camp David with Jimmy Carter, to sign anything to get out, this example does show that persistence is a characteristic that can pay big dividends in negotiations. People often need time to consider a new proposal. Their initial rejection may be more of a reaction than a firm position. They may need time to consider your proposal and adjust their thinking. If you give them enough time and information, their initial "no" may well change to "yes."

Knowing the Other Side

Knowledge is power.

Sir Francis Bacon

Obtaining and evaluating as much information as you can about the other side is a key step in negotiating success. All too often, people fail to realize the need for or value of specific information until they are in the middle of a negotiation—and then it is usually too late to get it. Information on their contractual terms, priorities, time pressures they face, their fiscal condition, patterns of past behavior, and any real limits that exist for them will be very useful. We've been in a negotiation where a public government client stated that his agency pays no more than 10-percent profit. Fortunately, we had reviewed contracts that they had signed with other consultants (which were all public records). We pulled one of those out that had a 12-percent profit, quickly ending the discussion on that point. We've had a purchasing agent tell us of a standard technique that he uses. He'll state that the maximum hourly rate he's prepared to pay is $100/hour, although his agency has paid more than that on many occasions. He says he is amazed that nearly everyone he negotiates with accepts his statement without challenge or knowledge that his agency will pay more. Always test these statements, which are attempts to establish the power of legitimacy of their position.

There are several ways to get information in advance of the negotiations. The most direct way is to ask questions of the other side. Don't assume that asking questions before negotiations will be pointless. Initiate contacts between your staff and their staff at the working level. People like to talk about themselves and their work. We suggest you review the portion of Chapter 2 dealing with open-ended questions. You may get answers far more useful than you ever anticipated. Even if the content of the answer doesn't tell you much, the manner in which the question is answered can be useful. This should be done in advance because once the negotiations start, the other side will assuredly clam up. For example, you can ask for past contracts to use as an example in drafting yours. If a public agency is involved, you can ask for copies of the minutes of their past council or board meetings. From these, you can find out who else they have recently signed contracts with and any details of the contract discussed in the meeting. Contacting others that have negotiated with the same customer can also provide information. If they are a private organization, ask their competitors for information on them. If you are buying a telephone system for your office, for example, it will be very useful for you to have cost data on competitive systems before entering negotiations. Sources of financial information, such as credit reports or Dunn and Brad-street reports, are also useful. Company biographies are available from Moody's and Standard and Poor's. Newspaper files, annual reports, stock reports, and litigation reports (which are commercially available) are also potential sources. Talking with people that have been involved in litigation with the other side can provide a wealth of information on how they behave in negotiations.

Set Goals

The only real risk is the risk of thinking too small.

Francis Moore Lappe

There is no way to determine the success or failure of your negotiations if you don't have a clear definition of your goals. Before beginning the negotiations, ask yourself the general questions:

- What do you want from this negotiation?
- What do you expect?

In answering these questions, you must consider the risks that you are willing to take. You need to look rationally at the potential benefits of a position to see if they are worth the risks of failure. View the goals of the immediate negotiation in the context of your personal long-term goals (see Chapter 5 for a discussion of goal setting), as well as your organization's long-term goals.

The negotiation process itself will provide feedback on your goals. Each proposal, concession, success, failure, and remark will cause each party's goals to shift up or down. Generally, people who set higher goals and make a personal commitment to those goals will do better than those that set lower goals. The goal becomes a firm personal intention, and once you are clear on your intention, the means to achieve it will often develop. As discussed in Chapter 5, positive visualization of achieving your goal can assist in achieving it.

The answers to the questions of what you want from the negotiation and what you expect can be quite different. Typically, people will set higher and more unrealistic goals for what they want as compared to what they expect. For example, you may want an 18-percent profit from the project you are discussing but expect, based on the historical practices of the customer or market involved, to get only 12 percent. There will be a higher level of personal commitment to what you expect than to what you want or wish for.

Constraints are not necessarily negative. They may force you to try avenues you would have ignored.

Cesar Pelli

Certainly, as part of your goal setting process, you need to consider at what point you'll reject the project. In the above example, if the negotiations appear stuck at a 10-percent profit level, you may ask for time to reconsider whether or not to proceed, or, perhaps, terminate the negotiations. However, there are some cautions about basing your willingness to continue negotiations on a "quit point" such as the 10-percent profit. It can preclude you from making adjustments that might become apparent if you'd continue to negotiate. You might have learned that the other side was willing to commit to two more projects with you. By focusing in on one aspect (i.e., profit level), you may inhibit the chances to develop an attractive overall package.

It's prudent to develop a list of items in advance that you are willing to trade for other more favorable terms. For example, you could decide to give up a markup of 5 percent on subcontractors if you can get an overall profit of 14 percent rather than 12 percent. In the real world, there are usually a number of interrelated factors that shape the goals of an individual negotiation. For example, your organization's goals may include items such as 10-percent profit, 15-percent growth per year, and expansion into a new market area. If you're negotiating the first project of its type needed to break into the new market area, you are probably willing to accept, if necessary, a profit less than your long-term goal of 10 percent in order to enter the new market. However, you should have a clear idea of the tradeoffs between profit and other contract terms that you are willing to accept before beginning the negotiations, to protect against impulsive agreement just to secure a strongly desired type of project. Closely related to setting goals is having a clear picture of alternatives should the negotiation fail. Defining such alternatives will give you good insights into the reasonableness of the goals you are considering.

If you lower your standards, you deserve everything you get.

Thomas Melohn

Developing Alternatives

Necessity never made a good bargain.

Benjamin Franklin

When entering any negotiation, there is the possibility that you will not reach a prudent agreement. What will you do in that event? Unfortunately, many people don't consider alternatives until the negotiations have failed or are about to fail, or until they've accepted a bad alternative because they believed they had no better choice. The best time to define alternatives is prior to the negotiation, when you can calmly and deliberately consider them—not in the pressure experienced in the midst of negotiations. Without a good alternative, you can be too committed to reaching an agreement, regardless of its terms. Consider how you'd feel about negotiating job salary and fringe benefits if you're in an interview and have no other job offers, compared with how you'd feel if you had another solid offer.

Try to arrange your life in such a way that you can afford to be disinterested. It is the most expensive of all luxuries, and the one best worth having.

W.R. Inge

Knowing your alternatives is very helpful in setting your goals. You can guard against setting goals that are not in your best interest. For the alternative to be effective, you need to know that it is real and what it looks like. Develop a list of alternatives (if you're part of a team, brainstorm with them), select the most attractive, and develop it so that it is well defined and real. If you're looking for new office space, you might first define your goal as: 5,000 square feet, within 10 minutes of the airport, all space on one floor, free parking, no more than $15/sq. ft./year including all utilities. You may discover one building meets all criteria but the costs are $17/sq. ft./year. Using an alternative location goal of 15 minutes from the airport, you find two other possibilities that meet your cost criterion. You've inspected them, found them satisfactory, and are willing to accept the location if you can't negotiate an acceptable price at the preferred location. You can now negotiate the price on the $17 space from a position of greater power. Without the alternatives, you would have been more inclined to accept the $17 because you didn't know if there was an acceptable alternative.

It is also useful to evaluate alternatives that the other side may have if an agreement isn't negotiated. They may be overly optimistic about their alternatives if they fail to negotiate an agreement with you. If so, you'll want to do what you can to lower their expectations. In negotiating on a project, our potential out-of-town client said that perhaps they should turn to an alternative, local consultant. In this case, we had information on the local firm. We knew that we had some specialized expertise that they did not. We pointed out the expertise needed to do the client's project effectively. We then suggested that if they felt they could get it from the local firm, perhaps they should because it would be more convenient for them. The client hired us. The more you know about the other side's alternatives, the better you can judge your position in the negotiating process.

Identify and Test Your Assumptions

Don't always assume that the other fellow has intelligence equal to yours. He may have more.

Terry Thomas

A lecturer at one seminar we attended wrote the word "assume" on the blackboard as "ass-u-me" saying that to assume can make an ass of you and me. Identify and test any assumptions you have about your case and the other side before you begin to negotiate. Check carefully how your assumptions affect your costs and schedule, such as the rate of inflation or periodic salary increases for a long duration project or the amount of time allowed for your customer to review your report before providing comments. Watch for an assumption that the project will run continuously and smoothly. Starting and stopping can be expensive. People often limit what they can get from negotiating by assuming limits that the other side won't exceed (e.g., they don't have that much funding, they'll never pay more than $100/hour).

It is important to identify any assumptions you've made and recognize that they are guesses, not facts, that could limit your thinking. Problems arise when you act as though the assumption was an absolute fact. You may not even be aware of the assumption. One demonstration we've seen illustrates this well. The speaker held up what appeared to be a standard, hexagonal wooden pencil. The audience was asked to call out facts about the object, being specifically cautioned to avoid assumptions. The responses included it was made of wood, had lead in the center, could be used to write and so forth. The speaker then bent the object which was made of rubber, showing that all of the "facts" were really assumptions. You can easily be unaware of your own assumptions! It is useful to identify assumptions that the other side will make about you, and to recognize that identifying assumptions—yours and theirs—will take a concentrated effort.

Develop Strategy and Tactics

Strategy deals with long-range goals and values. Tactics deal with the procedures and approaches to be used in the short term in a negotiation. The approach used to develop the long-term goals and values discussed in Chapter 5 can be used to develop company goals as well. When evaluating a specific project negotiation relative to company goals, you need to address the question, "How does this project fit into my long-term business strategy?" The time to address this question is before the negotiation. There are no right terms or price for a project that doesn't fit into your long-term strategy. If the project fits into your long-term strategy, then it is time to plan your tactics and how to respond to the other side's tactics. A separate section of this chapter address tactics.

Agenda

An important step in the planning process is to develop an agenda. If you plan and control the agenda, you control the pace and structure of the negotiations. An agenda can expedite a quick decision or allow careful explanations. In preparing an agenda, there are several factors to consider: the order in which issues should be introduced, how best to introduce issues, scheduling breaks or time to allow you to think, order of speaking, eating times, and identification of individuals responsible for specific tasks.

Rehearse

Luck is being ready for the chance.

J. Frank Dobie

We've seen cases where groups rehearse a sales presentation until it is a smooth, refined product and then proceed to negotiations with a team that has never worked together and that has not rehearsed the negotiation. The results are usually poor and could be improved dramatically by rehearsing. A particularly effective approach is to have various members of your team play different roles that will occur in the negotiations. For example, you can play roles as yourself, the other side's chief negotiator, an advisor to you or the other side, or, preferably, all the roles. Use the information you gathered on the client to make the exercise realistic. Such role-playing will allow you to focus on elements that may well be overlooked if you only discussed the negotiations. It will sharpen your approach because it puts you in the other side's place. Different approaches can be tested by instructing one side (without the other knowing it) to make a radical change in position (i.e., disallow expenses or cut the allowable schedule in half) and evaluate how your team responds. Videotaping the rehearsal sessions will allow you and others to analyze the approach and the tactics used and to develop appropriate improvements.

Tactics

There are a wide range of tactics that you may use or may encounter. Before examining some specific tactics, here are a few general guidelines:

- Be sure that *you* do the writing of any agreement that results. By doing so, the agreement will be in your terms.

A verbal agreement isn't worth the paper it's written on.

Samuel Goldwyn

- Start negotiations with easily settled issues rather than difficult ones.
- Presenting both sides of an issue will generally be more effective than presenting only one. "I understand that you have a severe deadline and a limited budget, and the resources required to meet your date are going to be costly."
- Where there is good news and bad news related to an issue, give the good news first.

- Tie controversial issues to issues where agreement can be reached to improve chances for success.
- Stress similarities of positions rather than differences.
- Use frequent perception checking by repeating the other side's statements to them for confirmation.
- When presenting the pluses and minuses of an issue, present those aspects that favor your position last.
- Approaches that satisfy the basic needs of the other side are most effective; any that threaten these needs will be rejected. Show the other side how your proposals meet their needs, even if they may not have expressed these needs to you.
- Explicitly state any conclusions you feel have been agreed to before moving onto another issue.
- Because most concessions occur at or beyond the originally perceived deadlines, be patient. You usually can't achieve the best outcomes quickly. With the passage of time, circumstances often change. Don't hesitate to go back to an issue if one comes up that affects it.

A man pretty much always refuses another man's first offer, no matter what it is.

Mark Twain

- Never begin the negotiations until you have developed your alternatives.
- Don't take anything to the negotiation session that you don't want the other side to see. In one negotiation session, the other side showed us a detailed task list. They said that they wanted us to prepare a labor estimate for each task, to compare with their independent estimate, which they were not willing to share with us. However, laying on top of a stack of their papers was their independent estimate! One quick glance gave us their idea of an appropriate level of effort.
- Don't take risks out of ego, pride, impatience, or a desire to get it over with.
- Take moderate risks, based on solid information, and ones that you can afford without being uptight about the results.

Before you can hit the jackpot, you have to put a coin in the machine.

Flip Wilson

- Without being pretentious, establish your credentials and expertise early in the negotiations so that it is clear to the other side that you actually have the knowledge needed to deal credibly with the issues.
- If you have an emotional issue or one that can be quantified (price, interest rate, profit, salary) raise it at the end of a negotiation, after the other side has made an investment of their time and energy. If it comes up early, discuss it briefly and defer resolution until later by saying, "I'd like to come back to that later after we discuss a couple of other items." People will be more flexible after their time is

invested in the process. The harder they have to work for it, the more they will want to make it work.

- Ask some questions (intelligent questions so you don't destroy your credibility) that you already know the answer to. The answers will tell you about the credibility of the other side.
- Don't rush to present your position. Always let the other side disclose their's first.

Success is more than knowing how. It's knowing when.

Advertisement for hair-growth formula, Rogaine

- Activate your ears and disengage your mouth. Everyone, including the other side, likes to be heard. You may be surprised by what you can learn or gain by just letting the other side talk.

Patience is a most necessary quality for business; many a man would rather you heard his story than grant his request.

Lord Chesterfield

- Speak in a moderate, controlled manner because a belligerent, loud, overbearing manner will change no one's mind and will likely result in similar treatment from them.
- If it's a one-on-one session, match their tone and pace to ease establishing rapport.
- Give yourself time to think by scheduling breaks, or, if necessary, ask for a bathroom or meal break or for time to consult with another member of your team or group.
- Use questions instead of statements. Questions will generate answers, while statements will generate resistance. If they fail to give you a sufficient answer, wait. They will feel uncomfortable with the silence, especially if they have their own doubts about their answer. You don't have to be talking to be effectively negotiating!

No man ever listened himself out of a job.

Calvin Coolidge

- Present your reasoning before you present your proposal. If you present your proposal first, the reasons will look like rationalizations for your position.
- If the other side insists on using their standard, preprinted contract, don't be intimidated by the printed word. You can always cross out and initial changes. If they insist on using a purchase order with standard terms printed on the back, you can write "Terms of Agreement Govern" on the face of the purchase order.
- Develop a detailed project scope before discussing price. The other side will receive an education on the project scope in the process and develop an under-

standing of the related costs. Generally, a detailed scope will help you get more money for the project, as opposed to less.

- Don't discuss an issue unless you are prepared for it. Acknowledge that it is a point that needs to be discussed and that you'd like to come back to it later in the negotiation. Winging it can be disastrous.
- Anticipate potential difficulties that may arise during the course of the agreement and develop contract concepts to address them before the negotiations start. Including such terms does not indicate a lack of trust; they are simply a means to avoid later misunderstandings that could damage the relationship. For example, maybe there is a potential that the project may be interrupted for several months. In this case, a provision to adjust reimbursement terms, personnel assignments, and schedule should be included.
- Be civil. As we said earlier, attack the problem, not the other person. A personal attack will heighten resistance because now the other side will be protecting their self-worth as well as their interests.

Civility is not a sign of weakness.

John F. Kennedy

Now, let's turn our attention to some commonly used tactics. Even though there are some that we don't recommend you use, be prepared to recognize them and know how to respond to them.

Take It Or Leave It

Better a friendly refusal than an unwilling promise.

German Proverb

There may come a time in a negotiation where you've hit the limit of how far you'll go. In this case, you should clearly state your final position (in a manner more politely than saying "take it or leave it"). Such a statement should only come after extensive negotiations. The chances of success are better if the other side has an investment of time and energy. You can reduce hostility to such a position by providing a thorough explanation or supporting it with references or precedent. Let the other side express their reaction. By listening and asking follow-up questions, you may be able to find out what is holding up agreement.

If you're on the receiving end of "take it or leave it," there are several options. One is to ignore it, keep talking, and introduce other solutions. You can also respond in a manner that will give them an out by saying something like, "The maximum overhead you'd accept was 150 percent before we discussed the cost of the liability insurance that you require." Normally, the best approach is to change the nature of the project by adding or deleting scope or changing the schedule.

You've Got To Do Better

This is a time-honored tactic you may encounter. Whether or not the other side really believes it, they may try to put you on the defensive by saying, "You surely can do better than that." Ask them what specifically they have in mind and why they want the change. Ask for something in return like a longer project schedule, a quicker (or delayed) starting date, cash up front, or a commitment for future work. If that fails, try changing the scope to match the budget or schedule that they have in mind. Don't be intimidated by the tactic. Be prepared with a plan to modify the project as a response should this aggressive tactic be applied to you.

Limited Authority

You may work your way through a lengthy negotiation process. You believe that you have reached an agreement, only to have the other side say that they must take it to someone else for approval. You car buyers recognize this one, and we've seen it happen in negotiations in much more "professional" circumstances. To avoid this, ask at the start of the negotiations if the representative has the authority. If they refuse to have someone with authority in attendance, make it clear that you are reserving equal freedom to reconsider any agreement. If you fail to ask at the start and they unexpectedly announce that they must take the "agreement" to a higher authority for approval, say that you will also treat the agreement as a draft and that you'll be back with your changes as well.

There are circumstances where it may be in your interest to have limited authority at a negotiation. If so, you should make the limits of your authority clear at the start of the negotiation, to establish your credibility and a sense of fairness and openness. Limited authority reduces the personal pressure of a negotiation. You don't have to analyze the other side's positions and make binding decisions simultaneously. There may be less personal antagonism between the negotiators and attempts to manipulate or coerce each other because others are making the final decisions. We know of cases where negotiators with the authority to make decisions played a game of "limited authority" for these reasons. Although supposedly making phone calls to get a decision on a point of discussion, they were probably finalizing dinner plans. If you are dealing with a complex project, it may not be feasible to have all of the legal and technical expertise in the room. By having the need to discuss such issues with the experts, you have the opportunity to assess your position, think about alternatives, and, if necessary, say "no" gracefully.

Refusal To Negotiate

What happens when the other side refuses to negotiate? Such a position is usually a negotiating ploy to gain some concession just to start negotiations or to establish preconditions for negotiations. The first step is to ask them about the reasons for their position. Don't attack them personally for not negotiating, but rather attempt to learn their reasons for their position. Are they concerned that no agreement is possible; about being criticized for talking with you; about the effects on their own group if they enter negotiations? You can then suggest some options such as changing negotiators, using third parties, using a mediator, or exchanging of letters on each side's positions.

You can also ask them if they want you to set preconditions for negotiations. How would they react if others refuse to negotiate with them? By asking for the principles that justify their action, you may be able to cause them to rethink the reasonableness of their position.

Buy Now, Negotiate Later

Make your bargain before beginning to plough.

Arab Proverb

We have run into several occasions where our potential client wanted us to begin work now and negotiate later. For example, a developer may want you to begin work on a set of plans with the offer to negotiate a contract later. A service is worth much more before it is rendered as opposed to after it is rendered. As undesirable as it may be, you may find yourself in the position of having to respond to such a situation. There are a few steps you can take:

- Establish a deadline (a *firm* calendar date) by which you will stop work if you don't have a contract. To be effective, you must stop work on the date.
- Ask for a significant (25 to 50 percent) portion of your expected fee up front. This accomplishes two things: 1) tests the sincerity of the customer; and 2) protects you against the costs you will incur prior to a deadline for a contract. Do not work beyond costs covered by the up-front payment.
- Ask for higher than normal rates for your work because of the risk you are taking by proceeding without a contract.
- Establish a per diem rate until the final project cost is agreed upon.

Deadlines

He who controls time, controls the negotiation.

Jon Q. Reynolds

One of the most universal catalysts for action is a deadline. This is why retail stores have such great days on December 24. You may find a deadline to be a useful tool in negotiations—or you may find yourself facing a deadline imposed by the other side.

A deadline you establish can potentially cause movement on the other side because the consequences of not meeting the deadline are uncertain. If they accept the deadline, they are dealing with less uncertainty. They may be willing to accept terms more favorable to you rather than face uncertainty. Because most major concessions or movement in negotiations occur near deadlines, knowing any real deadlines faced by the other side can be a tremendous advantage to you. As their real deadlines near, the pressure on them to make decisions increases. When you make an offer to purchase a house, it is accompanied by a deadline. The deadline causes the seller to make a decision or a counteroffer—it produces movement.

When the other side places a deadline on you, analyze it rather than blindly accepting it. Ask what the certainty and extent of consequences are if you go beyond the deadline. For example, April 15 is a deadline that every U.S. taxpayer is aware of. What happens if you miss it? If you have a refund coming, nothing will happen other than you will receive your money later rather than sooner. If you owe them money, you can compare the costs of interests and penalties charged by the government with those charged by a bank. So long as you eventually file an honest return, you'll not be dragged off to prison. So you can decide to whom you'd rather pay the interest—the government or a bank. Certainly, a very high percentage of people accept the April 15 deadline without analyzing it. That's why there are always lines at the post office at 11 P.M. on April 15! Of course, in a negotiation, a deadline may be real and there may be serious consequences if it isn't met. We're only suggesting that you carefully analyze deadlines—either stated by the other side, or one you've placed on yourself—before deciding how to respond to them.

The "Nibble"

A "nibble" involves gnawing at the edges of an agreement to gain advantages in small bits. The most difficult nibble to deal with is that which occurs after the agreement has been signed—a phenomena also known as "scope creep." Nibbles also occur during negotiation after the major issues have been resolved. Let's say you've agreed on the major scope items—the extent of the project, who is to do what portions of the work, when it will be done, and so forth. Your major worry items have been resolved and you're starting to count this one as in the bag. The buyer may then try to nibble at your proposed profit by trying to reduce it from 15 percent to 14.5 percent; reduce your proposed overhead (trying to reduce it from 150 percent to 145 percent); cut your schedule from 48 weeks to 46 weeks; or get you to throw in some extra copies of the project report, evaluate one more alternative, or attend extra meetings. No one item amounts to much, but cumulatively they can be significant.

At least during negotiations, you have a clear "yes" or "no" opportunity. After the negotiation is complete, it becomes more difficult. The customer may ask you to attend an extra meeting or submit an extra progress report. A common reaction is that the item seems very small in the big picture and it isn't worth requesting more time or money. Again, cumulatively, the result can be significant. The best response, whether in negotiations or afterwards, is to let the other side know that you recognize the nibble by quoting a price and time for the extra item at the time the item is requested. Consistently following this practice will soon alert the other side that you are alert for scope creep and that you expect to be paid fairly for what you do. If you choose to give away something, such as extra copies of the project report, send an invoice for the extra copies with a "no charge" notation. Your customer will then be aware of the extras they are getting.

Good Guy/Bad Guy

You've probably seen this technique in its classic form in the movies. A suspect is questioned by a tough detective under bright lights for hours. Eventually, the tough detective leaves and is replaced by a friendly police officer who shuts off the light, offers the suspect a cigarette, and lets him relax. The suspect proceeds to tell the good guy everything he wants to know. The same technique is used in negotiations. One member of the other team makes harsh demands in an aggressive manner with little or no room allowed for discussion. When he or she eventually finishes, another member of the team, smiling, talks in a conciliatory, friendly manner. He or she seems so reasonable by comparison that you are eager to agree.

The first step is to recognize this tactic as an attempt at psychological manipulation. This reduces its chances for success. Next, pursue the justification for their positions, rather than attacking the bad guy or being swayed by the good guy. For example, you may have asked for nine months to do the project. The bad guy may have been insisting that the project be completed in six months (only an idiot would take longer, the company will go broke if it isn't, the competition could do it in five months, and so on). The good guy may ask the bad guy to consider seven months. Rather than jump on the offer of seven months, ask them to explain why six months is reasonable, why they feel it is consistent with the quality of service you want to provide, how they see the logistics working in such a time frame, and so on. As with many tactics, focusing on the merits of the issues rather than responding to the tactic is a positive response. If you know the other side uses this approach, take along your own "bad" guy. You can then confront the client with the tactic. Suggest that you could both save time by sending the bad guys home and negotiating without them.

Physical/Verbal Abuse

"Don't let them push your button." The one trying to get you angry wants to control you. If you meet a negative approach positively, you are not letting the climate get out of your hands.

Gerard Nierenberg

Abusive tactics are counterproductive in achieving agreements that will be the basis of a long-term relationship. Although we do not suggest that you use them, you may encounter others who will attempt to apply them to you. We once attended a negotiation as a consultant to the team representing a large western city in negotiation with a large power company. The city team was led by a nationally-known and widely respected attorney from Chicago. The opening words of the negotiation came from the power company president and were directed to the city manager, another member of our team. He said, "At least we didn't bring a whore from Chicago to this meeting." Both the manager and the attorney were experienced negotiators who recognized this as a verbal attack designed to knock them off balance emotionally. Neither responded, or even blinked. The tactic failed. Recognizing the tactic will do much to negate its effect. Calmly saying "I suppose your verbal abuse is meant to upset me" is effective in preventing its recurrence.

Whenever one large landlord we know had to negotiate with a financial institution for interest rates or terms on his real estate development loans, he would meet with the individual in his conference room. He would close the doors and smoke a very large cigar throughout the meeting, filling the conference room with cigar smoke. When meeting with potential tenants, however, the cigars were rarely (if ever) seen. If you're feeling under stress physically because the room is too smoky, too hot, too cold, or too small, the sun is in your eyes, the session is too long, or you need a meal break and none is offered, recognize that this could be a tactic to hurry you through negotiations. Identify the problem, raise it with the other side, and request a better physical environment for the negotiations.

Limited Funds

You may encounter a customer who says, "I love your project proposal but I only have $40,000 available. Your price of $50,000 is just more than I can handle. Can we go ahead with the funding that I have?" There are several things you can do to respond to this tactic. First, have alternatives defined that can reduce your costs by reducing your scope to be consistent with the funds. Offer to let the customer do some of the project work rather than you doing it with your staff (data collection, for example). Pursue a change in the payment terms, allowing payment over a longer period, for example.

THE NEGOTIATION SESSION

Earlier sections of the chapter deal with planning, strategies, and tactics that may be applied during the session. This section reviews the physical environment, steps that constitute a structure for the session, and how to handle concessions. Although we're talking about the negotiation "session," negotiation is not really an event, it's a process. It is the opportunity to act and behave so that you develop trust, acceptance, respect, and acceptance. It is more than applying the tactics we've talked about earlier or some of the details that follow. It involves relating to the other person by your approach, attitudes, methods, and the way you relate to their needs and feelings. The process itself can go a long way toward fulfilling the needs of both sides.

Environment

To provide an environment conducive to productive negotiations, there are several elements to consider:

- Size of room relative to number of participants.
- Access to telephones, copy machines, FAX machines.
- Noise from external sources.
- Control of interruptions.
- Control of light and temperature in the room.
- Availability of secretarial help.
- Nearby rooms for each group to caucus in, if needed.

If you find yourself in a situation where you are uncomfortable due to the physical conditions (such as temperature, light, noise) or without facilities you need (such as telephones, copiers, secretarial help), call the conditions to the attention of the other side and ask for a change before continuing.

Do not attempt to use physical conditions to put the other side at a disadvantage (e.g., put them in chairs facing the sun). They will soon realize what you are attempting. You will win neither their trust nor confidence by such tactics.

Steps

There are several steps that can occur in a negotiation session. Not all will necessarily be present, and they will not always be in the same order. Keeping these steps in mind may provide a structure useful to keep the session focused. They can include:

- **Introductions, initial remarks**—A review of the background leading to the negotiation may be useful.

- **Statement of the issues**—Clearly summarize the reasons for the negotiation. It's prudent to check the other side's perceptions of the reasons before beginning to negotiate. Also, getting agreement on a list of issues to be negotiated can be a valuable first step. It can get both sides looking to the end result so that you work together to find ways to get there. This is a much better start than comparing your way versus their way. It helps emphasize ends, not the means.
- **Ground rules**—Work out arrangements for support services, caucusing, hours, breaks, attendees.
- **Agenda**—Be sure that all of the items important to you are on the agenda and, if possible, in the order that you prefer.
- **Discussion**—This is the heart of the negotiation, where agreements are worked out and where you apply the planning, strategies, and tactics that we discussed earlier.
- **Conclusion**—It is vital that both sides agree to what they've agreed to so that there are no misunderstandings. Agreements on various issues may have occurred at different times in the negotiation. It is important that the agreements be reviewed so that both sides leave the session concurring.
- **Writing the agreement**—Usually, the agreement is written after rather than during the negotiation session. As discussed earlier, volunteer to write the draft—keep the power of the pen.
- **Review and finalize the agreement**—When both sides see the verbal agreements committed to written words, another round of negotiations often occurs. If the other side has drafted the agreement, it is only prudent to carefully review it. It is often beneficial to explain the implications of each contract section. This can develop trust and minimize future misunderstandings. Also, people not involved in the discussion may have had a hand in preparing the agreement, causing some inaccuracies. There is always the chance that the other side·may, when they draft the agreement, try to win back some points lost in earlier discussions. It is then that good notes of the agreements summarized during the conclusion of the negotiations will be valuable. If the other side comes back with a draft agreement that is substantially different than agreed to at the session (this has happened to us!), ask them if there has been a misunderstanding or if they have changed their minds and wish to withdraw. If they are looking for a way out, it is best to recognize it rather than berate them, attempting to force a solution.

Concessions

Let us never negotiate out of fear. But let us never fear to negotiate.

John F. Kennedy

No matter how solid your case or how careful your planning, it is likely that you will face the need to make a concession during the negotiation session (otherwise it would be better named a capitulation session). Here are a few basic things to keep in mind:

- In most negotiations, it is likely that the other side will start with their maximum position without a hint as to their minimum position. They do have room for concessions. Don't take their first offer without thoroughly exploring areas for concessions.
- In a similar manner, it is prudent to leave yourself some room for concessions by starting at a point above your minimum (but not so absurd as to create hostility). As discussed earlier, check your assumptions before negotiations. It is possible that you may set a maximum level that is still not high enough because you may not know the value that the other side places on your services. In order for them to be satisfied about the negotiation process, you need to leave room for concessions. How do you feel when someone accepts your first offer? You wish you would have offered less!
- Be patient. Let the other side make the first concession on major points. It's best not to make any concessions until you know all of the other side's positions and demands.
- Tell the other side that any interim concessions are tentative, pending agreement on all other terms. For example, it is reasonable that your concession to reduce project completion time is tentative until the question of overtime pay is agreed upon. It is the final agreement that counts, not interim concessions made along the way.
- Keep track of the number *and* value of concessions by each side. This is information that can be a useful tool in bargaining on a later issue. An equal split in concessions isn't your goal. The idea is to keep yours less than theirs.
- Be careful about the pattern of your concessions and observe that of the other side because it can lead to a conclusion about minimum positions. For example, if the proposed project price starts at $150,000 with a second offer of $120,000, it would be difficult to believe that the minimum position is $115,000. It would appear more likely to be lower, maybe under $100,000. If the second offer is $140,000, you'd conclude that the minimum is probably going to be well over $100,000. The increments of concession give strong clues as the limits of the negotiation. Make them small.

Pay careful attention to the increments of change in your opponent's demands. When the increments begin to decrease in size, your opponent is reaching his or her bargaining limit.

Jeff Furman

- Don't concede too much, too quickly. You'll raise the sights of the other side. You may actually reduce the chances of agreement because the other side raises its sights to unrealistic levels.
- Be able to support any concessions you make or that you ask for with objective, rational arguments, and/or data.
- Remember, you can say "no" to demands for unacceptable concessions. As discussed earlier, by defining your alternatives in advance, you'll be in a stronger position to accept or reject concessions.

Learn to say "NO"; it will be of more use to you than to be able to read Latin.

Charles Haddon Spurgeon

Breaking a Stalemate

You may reach a point in the negotiations where you and the other side have both invested a substantial amount of time and energy and cannot reach an agreement. It may be that the true cause of the stalemate is not a technical issue, but is rather a human issue. The other negotiator may be afraid to decide, may be worried about the reaction of his or her boss, may have a divided negotiating team, or may be worried about losing face by backing off a strongly voiced position. Here are some ideas to get the discussion going that provide some room for the other side to address their human factors:

- Change the negotiators.
- Change the project scope. Delete or add an element. Now you're talking about a new project and previous positions may no longer apply.
- Change location of negotiations.
- Change the schedule.
- Take it to a higher authority on each side.
- Change the payment terms, contract language, or type of contract.
- Throw out some new options.
- Change the specifications for the completed project.
- Change the project manager or project team members.
- Bring in a team member or expert for new ideas on one issue.
- Recess to gather more information on the difficult points.
- Creatively look for an objective, fair procedure to resolve the difference. Consider the oft cited example of two children deadlocked over how to divide a piece of cake. A fair procedure is to have one cut and the other choose. Neither can complain about an unfair division. If you're being asked to sign a subcontract and are stalemated over terms, ask the prime contractor to provide a copy of the terms of the agreement when they were last a subcontractor.
- Call in a third party. They can be used to define the issues and facilitate

communication so that all viewpoints are clearly presented on each issue, foster more respect and cooperation, and, in some cases, offer creative solutions or alternatives to consider.

Involving a third party to resolve a stalemate has some disadvantages. It is an acknowledgment that the two sides have not handled their differences. As a result, the motivation and sense of responsibility of the negotiators can be subtly damaged and the long-term relationship between the two parties can be downgraded. This rather subtle long-term potential effect must be weighed against the shorter term advantage of potentially resolving the stalemate. The third party may also force an agreement not in the best interests of one or both parties. After all, the third party doesn't have an investment in the eventual outcome of the project.

Closing the Negotiations

Hopefully, you will reach a point where you are satisfied with the results. You have agreed on the critical project terms of:

- Scope.
- Deliverable products.
- Responsibilities of both parties—and the individuals responsible for meeting these responsibilities.
- Liability and indemnification considerations.
- Schedule, including a start date and key milestones.
- Price.
- Payment terms.
- Personnel assignments, which cannot be changed without client approval.
- Procedures for project changes and related modifications in price or schedule.
- Any applicable regulations or codes.

When you have the verbal agreement you want, take a positive approach by moving on to formalizing the agreement. Ask when they would like to have a draft agreement. Offer to draft a written agreement. Project a positive attitude by praising the joint efforts made to achieve a solid basis for the project and by expressing your excitement and eagerness to get on with the work. If you have applied the cooperative approach described in this chapter, the issues will have been resolved while building positive personal relationships.

When you're done with the process, do a critique of what worked for you and what didn't. Reviewing a file of these critiques before starting your next negotiation can be a valuable exercise!

Keep a file of your negotiation planning, tactics, client information, and so forth, for future negotiations and for training purposes. These files can be useful in future role-plays and rehearsals. They provide a useful comparison of how the rehearsing teams do, to the actual past results.

Experience enables you to recognize a mistake when you make it again.

Franklin P. Jones

PREPARATION CHECKLIST

Figure 4-1 presents a checklist that is useful in preparing for negotiations.

(Attach supplemental sheets as needed for answers)

Our Negotiator(s)	
Use individual or team	Individual _____ Team _____
Individual selected	Name
Team members	Names Roles
Who commits our side	Name
Information on other side	
Review past contracts for: - Allowable profits - Allowable overheads - Allowable hourly rates - Contract language	
Key Issues - List the three most important project elements to them in descending order and why.	1) 2) 3)
Review financial statements	
Who is their negotiator(s)?	
What is his/her relation to project?	
What is most important to their negotiator?	
What alternatives to this project and/or us are available to them?	
What are funding considerations they face?	
Why has authority to commit them? - Will they be at the negotiation?	Yes _____ No _____
Key Project Considerations	
Our understanding of project objective	
Scope items essential to us	
What are deliverable items?	
What factors affect schedule?	
What are potential areas of liability?	

FIGURE 4-1   Checklist—Preparing for Negotiations

Our Positions	

What do we want?

Profit	Opening Position_____	Goal _____	Min. Acceptable _____
Total Price	Opening Position_____	Goal _____	Min. Acceptable _____
Schedule	Opening Position_____	Goal _____	Min. Acceptable _____
Hourly Rates	Opening Position_____	Goal _____	Min. Acceptable _____
Pymnt Terms	Opening Position_____	Goal _____	Min. Acceptable _____

Key contract language - Identify issues (i.e., indemnification, key personnel, project interruptions, penalties, etc.)	
Contract type sought/desired (lump sum, cost plus fixed fee, per diem, etc.)	
What alternatives do we have?	
Preparation for Negotiating Session	
Agenda prepared	
Rehearsal of negotiations	
Actual negotiation environment - Room size adequate? - Access to telephones? - Access to photocopier? - Access to FAX machine? - Access to secretarial help?	
Can noise, light, temperature be controlled?	
Nearby rooms to caucus in?	

FIGURE 4-1 (Continued)

References

1. Fisher, R., and W. Ury. 1981. *Getting to YES. Negotiating Agreement Without Giving In.* New York: Penguin Group.
2. Cohen, H. 1982. *You Can Negotiate Anything.* New York: Bantam Books.
3. Nievenberg, G.I. 1981. *The Art of Negotiating.* New York: Pocket Books.
4. Stasiowski, F.A. 1985. *Negotiating Higher Design Fees.* New York: Whitney Library of Design.
5. Karrass, C.L. 1970. *The Negotiating Game.* New York: Thomas Y. Crowell.
6. Karrass, C.L. 1974. *Give and Take.* New York: Thomas Y. Crowell.
7. Shea, G.F. 1983. *Creative Negotiating.* Boston: CBI Publishing Co.
8. Rodman, L.C., and M. Hensey. 1991. Project manager's role in contract negotiations. American Society of Civil Engineers, *Journal of Management in Engineering*, p. 5. January 1991.
9. Karrass, C.L. 1989. *Effective Negotiating—Workbook and Discussion Guide.* Santa Monica, California: Karrass.

5

Increasing Personal Effectiveness

If your aim is management, it must be self-management first.

E.B. Osburn

It shouldn't come as a surprise that the best project managers we've seen are also very effective at managing their own time. The best project managers are not workaholics that are at the office night and day and weekends, striving for the attention associated with constant activity.

A great many people have asked how I manage to get so much work done and still keep looking so dissipated.

Robert Benchley

Workaholics are addicted to activity, while the best managers are committed to achieving results. Some of the best managers that we've worked with are around the office amazingly close to 40 hours per week and lead well-adjusted, rewarding lives. Your own personal effectiveness depends on what you do with the finite amount of time that you have. No one has more than 1,440 minutes per day, yet some people are clearly more effective than others. There is a direct correlation between how well you manage yourself and how well you manage your projects. There are some simple essentials that you can use to improve your personal effectiveness.

SET LONG-TERM GOALS

"Cheshire Puss—would you tell me please which way I ought to go from here?"
"That depends a good deal on where you want to get to," said the cat.

"I don't much care where . . ." said Alice.
"Then it doesn't matter which way you go," said the cat.

Lewis Caroll, *Alice in Wonderland*

The way you use your time is the route you travel to your goals. We bet you didn't hop into your car this morning and begin to drive aimlessly without knowing exactly where you were going. Yet, how many of us start the day without a clear picture of what we want to accomplish in the next eight hours, or in the next year?

The key to personal effectiveness is defining goals—both long- and short-term. Long term goals are the place to start—you need to know whether you're going to New York or Los Angeles before you decide which way to turn onto the freeway.

It takes time to save time.

Joe Taylor

Most people resist long-term planning because it takes thought and commitment. The crisis of the day seems to preclude the time needed to plan. Management literature is full of studies that show that every hour spent in planning returns a dividend of many hours, not to mention long-term satisfaction; yet it is common to procrastinate about planning. Instead, we attend to problems that planning might well have eliminated. There is an auto maintenance television advertisement that ends with "You can pay me now or pay me later." The same can be said for planning. The emotional energy and time to redirect your energy once you've discovered you've headed down the wrong route can extract a great price that can be avoided by planning.

Ours is a world where people don't know what they want and are willing to go through hell to get it.

Don Marquis

It is important to consider the nature of "goals" before talking about setting them. Goals can obscure a long-term vision or mission. For example, a goal of driving 500 miles per day isn't useful unless it is tied to the vision of driving from New York to Los Angeles. Some people get caught up in checking tasks off their daily to-do lists while forgetting where these tasks were intended to take them. They are always busy but don't accomplish much. Daily completion of the tasks may make you feel good in the short term, but unless they are tied to a long-term vision, your satisfaction and performance will eventually decline. When we're referring to "goals," we're referring to those associated with a long-term vision of what you want to accomplish—what you want your life to look like.

I always wanted to be somebody, but I should have been more specific.

Lily Tomlin and Jane Wagner

There are a lot of ways to go about setting long-term goals. We'd like to tell you about one that we've found to be especially effective. You'll need a few sheets of paper for this exercise.

There is much to be gained by writing down your long-term goals. You will be much more specific than you are in occasional random thoughts, such as I'd like to travel more, retire at age 40, and so on. By writing them, you may well discover important goals that you hadn't seriously considered before. Others may drop in importance or off the list totally. Written goals can be periodically reviewed, considered, and revised and can be referred to in setting daily, weekly, or monthly strategies.

On the first sheet of paper, list those areas of your life where you're expending or want to expend energy, such as career, family, various relationships, community, health and exercise, finances, education, spiritual, and so on. Take your time. Be sure you've done the best you can at identifying them, recognizing that the list can be expanded later.

Above all, challenge yourself. You may well surprise yourself at what strengths you have, what you can accomplish.

Cecile M. Springer

The next step is to take each of these areas and visualize what it would look like (the "ideal scene") if it were ideal. What's the experience that you're looking for? Use a separate sheet of paper for each area you've identified. One approach is called "mind mapping." Place your goal in the middle of the page and simply let yourself brainstorm as to how achievement of your goal would look. Figure 5-1 is an ideal scene for being an effective project manager. Write down the ideal scene in each area of your life. Find a quiet spot and allow plenty of time. Devote as much of the day as the process may take—a day when you are away from the distractions and pressures of the office. *Don't* think about *how* you're going to reach the ideal, but visualize what it will look like when you have reached it.

You always move toward your dominant thought.

Dennis Waitley, *Seeds of Greatness*

Clearly defining your goals allows you to focus your thoughts and energy in a positive manner. Indeed, your energy will follow your thoughts. Lou Holz, a very successful football coach and motivational speaker, offers an example. At age 30, he wrote down a list of 107 goals he wanted to achieve in his lifetime. They ranged from dinner at the White House to owning a 1949 Chevrolet. By the time he was 52, he had achieved 86 of his goals. Would he have accomplished as many without committing them to writing?

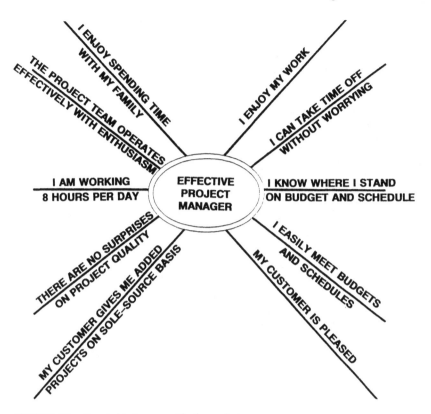

FIGURE 5-1   Sample ideal scene—effective project manager.

*Destiny is not a matter of chance, it is a matter of choice; it is not a thing to be
waited for, it is a thing to be achieved.*

William Jennings Bryan

Positive visualization of your goals as if already achieved can be very powerful. It
seems that many engineers and technical people have a skepticism of such non-tangible
concepts and are much more comfortable with immediately establishing tasks and
schedules to achieve a result. As you'll see from the rest of this chapter, we are not
discounting the value of specific strategies and plans. We are suggesting that you can
achieve surprising results by positively visualizing yourself enjoying the successful
accomplishment of your goals. Athletes have used this concept by visualizing themselves executing the perfect golf swing or running the perfect race over and over in their
minds. Jack Nicklaus explained the technique in his book, *Golf My Way*:

First I "see" the ball where I want it to finish, nice and white and sitting up high on the
bright green grass. Then the scene quickly changes, and I "see" the ball going there:

it's path, trajectory, and shape, even its behavior on landing. Then there's a sort of fade-out, and the next scene shows me making the kind of swing that will turn the previous images into reality.

Take some time each day to see yourself achieving your goals. Don't cloud this time with thoughts of how you're going to reach the goal or all the difficulties you need to overcome to get there. Focus only on seeing yourself enjoying the ideal scene you've developed in the goal setting process.

Clinical psychologists have found that the most effective way to do this is to develop positive self-statements or engage in positive self-talk. These statements work best if:

- You use personal pronouns, such as "I" and "my."
- They are in the present tense.
- They are concise.
- They are directed toward what you seek rather than what you want to avoid.
- They are aimed at improvement rather than perfection.

For example, an ineffective statement for a health goal would be "I will lose ten pounds"; an effective statement would be "I am maintaining my ideal weight of 155 pounds and enjoying exercising and feeling energized every day."

Man is not the creature of circumstances; circumstances are the creature of man.

Benjamin Disraeli

You may be amazed at how the means to achieve your goals appear when your intentions are clear and strong. About 20 years ago, I (Gordon) had a vision of leaving a job at a large engineering firm to start a consulting firm. The experience of starting up a new branch office for the large firm had been the catalyst for the vision, realizing that the experience gained was much the same as starting a new firm. Still being relatively young with a growing family, having a few months of cash flow during startup was a major problem—enough of a problem that the concept remained only a vision, but a strong one. I continued to visualize myself as a successful independent consultant, even though I could not identify any practical means to get from where I was to the goal. I was not out looking for loans or trying to line up clients. I was devoting my conscious energy to doing my job while maintaining a positive mental picture of eventually being a successful independent consultant.

Within about four months of focusing on this vision, I received a phone call from a venture capital group looking for investment opportunities in my field of technical expertise. I inquired as to their interest in investing in a new consulting firm. Although they weren't interested in investing, they were interested in hiring an individual consultant to evaluate investment opportunities. For an exclusive arrangement, they were willing to pay for a substantial effort *in advance*! The cash flow problem was solved and I started my own firm. The means to the goal appeared. In the process of positive visualization, you actually *see and experience* yourself at a specific goal and you maintain a regularly expressed and sustained focus on this vision. Maybe you're skeptical of a connection between the positive visualization and the result. Maybe it was a coincidence. However, I've since had many similar experiences. Now even this engineer, steeped in the scientific method, is convinced that there is something to the theory that the means will become evident if the intention is clear. Maybe it will work for you. Check it out. Define your long-term goals. Visualize achieving them. Check the results.

May I ask you a highly personal question? It's what life does all the time.

Kurt Vonnegut

An effective and interesting way to check the results is to periodically review and question your long-term goals (the ideal scenes you defined). See how many you've achieved. Examine the ones you haven't. Are they still valid goals? Drop them or modify them if they aren't. If they are, focus your energy on them in the next few months and check the results.

ESTABLISHING STRATEGIES AND SHORT-TERM GOALS

Decide what you want, decide what you are willing to exchange for it. Establish your priorities and go to work.

H.L. Hunt

Your long-term goals are the framework for strategies and short-term goals. If you're driving from New York to California, your strategy is to move westward. There may be some detours and delays on the journey, but as you start each day, there is a clear sense of direction—west. Your long-term goals give you that same sense of direction.

The goal setting and monitoring process is analogous to an auto-pilot system in a boat. Once the direction is set, the system monitors feedback signals from the compass to adjust your course. Your boat may be on the precise course a very small percent of the time, but by integrating the feedback into action, you'll arrive at your destination. Your mind can work in the same way. Once your goal is set, the world will offer you plenty of feedback so that you can adjust your course. When Colonel Sanders started selling chicken in the late 1950s, he was 65 years old. His goal was to make $1,000 per month. The world gave him a lot of favorable feedback. He adjusted his course, franchised Kentucky Fried Chicken restaurants throughout the world, and made a fortune.

The best thing about the future is that it comes only one day at a time.

Abraham Lincoln

If you attempt to develop a single plan to accomplish your long-term goals, you will feel overwhelmed. We've heard this described as the "salami syndrome." Eating an entire salami in one piece is overwhelming and unrealistic and would probably give you indigestion. The best approach is to slice the salami into digestible bits. We've found that the same concept applies well to achieving your long-term goals. Our experience is that it is effective to first slice goals into pieces called monthly strategies and then to cut them further into digestible bits called weekly strategies.

The key is not to prioritize what's on your schedule, but to schedule your priorities.

Stephen R. Covey

Take a few quiet minutes once each month to write down your strategies for that month. What steps toward the long-term goals do I want to achieve this month? Once a week, write down your weekly strategies to achieve your monthly goals. A weekly focus is better than a daily plan. Daily plans tend to get bogged down in busywork. What do I want to achieve toward the monthly strategies this week?

Sunday evening may be an effective time to define your weekly action steps because it provides some time for the daily pressures of the work week to subside. You may be fresher, allowing a better perspective. You may find Friday evenings to be better because the events of the past week are fresh in your mind, helping you adjust for the next week. Try some various times until you find the best for you. This process takes surprisingly little time—usually less than 30 minutes—and will pay major dividends throughout the month and week. Figure 5-2 illustrates the process of going from a vision to reality.

Here are some keys to keep in mind when setting goals:
- Be sure they align with your ideal scenes.
- Give them realistic time frames.
- Write them down.
- Break them into achievable steps.
- Be flexible, as conditions may change.
- Ask for assistance if you need it.
- Be willing to commit to achieving the goal in terms of energy and time.
- Be sure it's your goal—not one you adopt to please someone else.

Ideals are like stars; you will not succeed in touching them with your hands. But like the seafaring man on the desert of waters, you choose them as your guides, and following them you will reach your destiny.

Carl Schurz

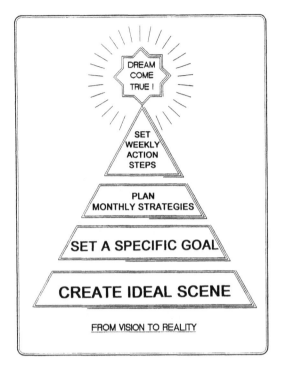

FIGURE   5-2

Later in this chapter, we'll talk about calendars and planners. Many of the available planners have pages (or additional pages can readily be added to the loose-leaf planners) that can be used to write monthly and weekly strategies. We urge you to do so! It is then convenient to refer to your strategies as you prepare your daily "to-do" lists.

PERSONAL STRATEGIES AND THE "TO-DO" LIST

You've got to be careful if you don't know where you're going, because you might not get there.

Yogi Berra

We find that most people have used a to-do list at work or in their personal affairs or both. We also find that far fewer people use a to-do list as an effective means of converting weekly strategies to daily actions, to maximize their personal efficiencies and as an integral part of moving toward their long-term goals. The to-do list involves listing on a piece of paper what tasks you want to do during the day. As the tasks are done, they are crossed off. As new tasks arise, they are added to the list. Pretty simple concept! There are some keys to making the list work efficiently.

- Use a single piece of paper each day for your list, preferably a page in your daily planner. Do not jot down items on random slips of paper as ideas occur during the day—the chances of loss are too great.
- Keep your list in a visible location throughout the day. *Do not* put it in a drawer or file!
- The basic process of planning is once again a key. Take a few minutes at the start of the day to prepare your list. Review long-term goals and your weekly strategies. List tasks related to them.
- Review your list from the day before and enter any items that still need doing.
- Give thought to what must be accomplished to meet upcoming deadlines.
- Put your personal and work items on the same list to give you the overall picture for the day.
- Supplement your to-do list with a calendar. The to-do list has no time scale so you can't keep track of when you did a task, which may be important to know later.

Who begins too much accomplishes little.

German Proverb

It is possible and even likely that your list will have more tasks on it than can be achieved during the day. Those that get the most out of a to-do list know that the key is *prioritization* of the items. Every discussion on prioritization that we've seen describes the Pareto principle—and for good reason! Vilfredo Pareto, a nineteenth-century Italian economist, noted that the most significant items in a group are normally a relatively small portion of the total items in the group. The concept has often been called the 80/20 rule, as well. Everyday life provides ample proof of the validity—20 percent of the fishermen catch 80 percent of the fish; 80 percent of a consultant's business comes from 20 percent of the clients; 80 percent of the personnel problems come from 20 percent of the staff, and so forth. You'll also find that, typically, 80 percent of the value of your daily activities will come from 20 percent of the items on your to-do list. Most of the small, quick tasks that keep us busy fall into the trivial 80 percent. We tend to do them because they are quick and we can cross them off the list.

Rule #1. Don't sweat the small stuff.
Rule #2. It's all small stuff.

Anonymous

The key is to prioritize the items on your list so that your energy is focused on the 20 percent of the items where the value resides. For example, in the consulting business, 80 percent of new business typically comes from existing clients, and most of that from a few select clients. Clearly, activities that keep those key existing clients happy should be high priority items on any consultant's to-do list. However, we've seen many examples where the excitement of chasing a new client leapfrogs an item right over the top of servicing an existing client. Keep the Pareto principle in mind as you set your daily priorities!

The whole point about getting things done is knowing what to leave undone.

Stella, Lady Reading

There are many techniques for prioritizing the tasks—none is right or wrong, use what works for you. A commonly taught method is to break the tasks down into categories of A, B, and C priority items. Category A items are of the highest priority—the must-do items, those that have the most value. Finishing a report promised to a key client or purchasing a present for your wife before tonight's anniversary dinner could be A-items. Tasks related to your long-term goals are likely A-items as well. Using the earlier example career goal list, completing an abstract of a technical paper for submission to a national conference by the deadline would likely be an A-item. B-items are those that are of medium value— nice to do if the time is available. C-items are those of low value or low urgency. Items may switch categories from day to day because you must consider the best use of your time each day, in light of the circumstances of the day. Completing your tax return could move from a C-item on January 11 to an A-item if still undone on April 14. Items of low urgency in the C-item usually don't jump from a C priority to an A priority overnight, but rather gradually work their way from C to B to A.

While I am busy with little things, I am not required to do greater things.

St. Francis de Sules

A variation of the above approach is to further prioritize items within a given category (i.e., defining items A1, A2, etc.). In our experience, this is an unwarranted expenditure of energy. Generally, as the day progresses, the order of attacking the A-items becomes clear. The most effective people are those that consistently get a large number of A-items done per day without getting bogged down in deciding which A-item to do first.

Another approach is to list the items in order of priority, 1 through however many items you have. This has some problems. What do you do with an item that arises during the day? It isn't worth the effort to renumber them all. Categorizing it as an A, B, or C priority is simpler and more effective.

Another variation is to weight each item based on both short-term urgency and long-term value, to avoid becoming a slave to urgency. Such a rating might look like (1 = highest value or urgency) the following table.

Task	Long-Term Value	Short-Term Urgency	Total Rating
Prepare for planning meeting	1	2	3
Research item and send to client	3	1	4

Not completing your to-do list each day is not a sign of failure. Contemplatively reassess the priorities for the next day, based on today's results. You'll find the few minutes spent in doing so pays off handsomely. Take a critical look at C-items to be sure they are worthy of staying on the list.

EXPECT THE UNEXPECTED

Faced with the choice between changing one's mind and proving there is no need to do so, almost everyone gets busy on the proof.

John Kenneth Galbraith

It is rare that we can make it through the day without surprises, some of which may radically shift our priorities and necessitate a change in direction. Such events do not mean that our planning is at fault. Certainly, the Apollo space program flights to the moon were planned in meticulous detail. Yet, while on the way to the moon, the Apollo 13 spacecraft suffered an explosion in its command module liquid oxygen system, the primary source of air for breathing. The priorities suddenly switched to using the lunar module's system and to devising a manually controlled course for the trip back. Occasionally, your priorities can be suddenly and drastically altered during the day. If it is more than occasionally, it may be a symptom of poor planning—that is, fighting fires instead of preventing fires.

There can't be a crisis next week. My schedule is already full.

Henry Kissinger

As unexpected demands for your time arise, keep your balance. Calmly assess the priority relative to other items on your list and how it fits into your strategies. Take some time to think through your reaction. "Just say no" is advice that can apply to more than drug abuse. Be realistic about how much you can take on. By being honest with yourself, you're being honest with others. Both you and they will respect you. Maybe the unexpected item is a C or less. Don't let it throw you off the track. Other people may ask you to do a task. Your effectiveness can drop dramatically if you consistently drop your A tasks to do what is a C for you but an A for someone else. If that someone is your boss or a client, it may be tricky to sort out the priorities, especially when both tasks are ones they have requested. In this case, point out the conflict in priorities and ask them to determine which one takes top priority. If you have two managers both giving you urgent tasks, ask them to negotiate priorities between themselves, taking you out of the loop.

Another valuable technique is to avoid an immediate response to an unanticipated demand. For example, a telephone call requesting you to give a talk at next week's Rotary Club may have a lot of immediate appeal. You'd sure like the exposure to the local business community and you're flattered by the invitation. Have you ever immediately said "yes" and later—maybe only a few minutes later—realized that the time needed to prepare is going to conflict with some high-priority tasks? We have. You're then faced with the bad choice of backing out of the commitment you just gave or letting some high-priority task suffer. When the unexpected opportunity such as this arises, respond by saying you're interested, you need to check a couple of things, and you'll call back within the hour, or ask to sleep on it. It's more difficult to renegotiate after a commitment is given. Look over your strategies, your priority to-do items, and assess how the unexpected fits in before giving an answer. When reprioritizing, you need to consider the consequence of not getting a task done. Which will cause the most damage if not done? It then becomes a higher priority.

Allow for accidents. Allow for human nature, especially your own.

Arnold Bennett

Many people are great at including contingencies in their project budgets, yet often seem to allow zero contingency in scheduling their own time. For example, we've been involved in coaching presentations where those involved have allowed no contingency for things that could go wrong in the production schedule (a printer fails) or in their own time for rehearsals (a client calls for help in the only hour they allowed for rehearsing). Common responses are that I am too busy, I don't have any slack, or I can't allow any contingency. These are sure symptoms of bad planning and poor prioritization. You will eventually have an outright failure or less than optimum results. Expect the unexpected—allow contingency time in setting your personal schedules.

USE YOUR PRIME TIME WISELY

A man who dares to waste one hour of time has not discovered the value of life.

Charles Darwin

Are you a morning person or a night person? It seems that each of us has a time of day when we are at our physical or mental peak. Kerzner reports that when 300 project managers were asked to identify their prime time, the majority felt that 9 to 11 A.M. was their peak time with a secondary peak at 1 to 4 P.M. Many believed that they did their most effective writing in the morning and saved their reading for the afternoon. They also felt that their performance varied from day to day, with the peak performance days being Wednesdays and Thursdays. Recognize when your "prime time" occurs. Schedule the routine tasks for your off-peak part of the day or the week. Zealously protect your prime time. If it normally takes until 9 A.M. to be firing on all cylinders, use the first hours of the day for the high priority tasks that take the least creativity on your part. On the other hand, if you illuminate the early morning sky with your brilliance, consider tackling the tough tasks at 7 A.M.—a time when you may get a double benefit because you're less likely to be interrupted.

Lose an hour in the morning and you will be all day hunting for it.

Richard Whately

CONFRONTING PROCRASTINATION

Never confuse activity with accomplishment.

Benjamin Franklin

Have you ever found yourself very busy but not accomplishing the important things? Procrastination is likely the cause. A decision not to decide is procrastination. Failing to act on a planned task is procrastination. Although we haven't seen any statistics on the subject, procrastination surely ranks high on the hit parade of time wasters. The goal setting process we've been talking about is of no value unless you act.

Why do we procrastinate? The task appears overwhelming. The task appears unpleasant. We are afraid of failure or the unknown. You can probably add your own favorites to these causes of procrastination.

There is nothing so useless as doing efficiently that which should not be done at all.

Peter F. Drucker

If a high-priority task is remaining undone over a period of time or is consistently being delayed day after day, the first step is to determine if the task really belongs in the high-priority category. Maybe the task is the wrong one. Make a list of advantages and disadvantages for the task. Critically examine the balance between them. Is the task the right one? If the task is the right one, doing anything else is a waste of your time. A natural temptation is to do the easy tasks or those that take little time rather than begin the procrastinated task. Doing so is merely substituting motion for progress on what is really important. Sure, there is satisfaction gained from crossing off several tasks from your to-do list, even if they are Bs or Cs. We've known cases where people do a task not on their list, and then add it to their list so that they can enjoy crossing it off! Maybe it will only take five minutes to update your expense report, but if it isn't due for several days, it should remain a C task left for later. Doing the unimportant tasks diverts your focus from the important ones, even if not much time is required to do them.

I must have a prodigious quantity of mind: it takes me as much as a week sometimes to make it up.

Mark Twain

Here are a few steps to overcome procrastination:

- **To overcome overwhelm, slice the salami**—Break the overwhelming task into doable tasks. For example, the task of writing a book could be overwhelming in light of all the usual work assignments and day-to-day tasks. Rather than look at the task as "write a book," it can be cut into smaller, doable pieces, such as prepare an outline of one chapter per week, write one chapter a month, and so forth. Any task can be broken down into smaller pieces. Defining the smaller subtasks, an output (i.e., a draft chapter) for each subtask and a schedule for each is a simple but powerful tool in fighting overwhelm. It also increases the success rates. For example, in pitching horseshoes, it is a lot easier to make a 3-foot toss than a 15-foot toss. Your goal in dividing an overwhelming task is to set up a whole series of doable 3-foot tosses.

It's a cinch by the inch; hard by the yard.

Anonymous

- **Define a time for action**—Maybe there is a good reason you cannot act on the high-priority task (the chapter outline) now. For example, a commitment made to your boss or client to finish another task may be staring you in the face. Rather than fret about the task, make a commitment to yourself to begin the high-priority task at a specified time: work on the outline for Chapter 1 on Saturday from 9 A.M. to noon. Write the commitment on your calendar. Then don't worry about it until Saturday. Give it all the priority you would give a commitment you've made to someone else, or you will always end up last in line. You deserve to do things for yourself. A key to breaking procrastination is to get involved in the task at any level, and scheduling a definite time for action will begin the involvement.
- **Do something physical**—For example, put all the references for Chapter 1 on your desk. Seeing them there will keep your high-priority task in the forefront of your consciousness. As a next step, take everything else off your desk, other

than the references and a pad of paper. Change your location. Take the references and go to an empty conference room, or pack them up and go home to work on the outline. The change in environment and elimination of interruptions and distractions can combine to really focus your attention.

Even if you're on the right track, you'll get run over if you just sit there.

Will Rogers

- **Set intermediate goals**—Set goals such as completing one chapter outline in the next week. Reaching these goals provides positive reinforcement—progress is visible! They serve as reference points from where you began. For example, I (Gordon) decided to learn to swim at the age of 50. Fifty may be a young age for many activities, but it is a difficult age to learn to swim. Very few instructors were even willing to take on a student of this age. For someone who couldn't even float, learning to swim appeared overwhelming. As the lessons progressed, I began to set intermediate goals. There was a large bear painted on one wall, about one-third of the way down the pool. The goal became to be able to swim to the bear! Then the goal was to swim to the ladder two-thirds of the way down the pool. Then the entire length. Then laps. Now I'm learning to water ski. Although progress in any given week may not have seemed like much, looking at the achievement of the intermediate goals relative to the starting point provided stimulus to keep on going.
- **Assume success**—Dwelling on the possibility of failure can be a major contributor to procrastination. Recognize that you are fantasizing about failure—you haven't even begun the task. You can just as well choose to fantasize about a positive result. Visualize yourself enjoying the success of the completed task. The positive visualization process we talked about earlier can produce great results, including breaking procrastination.

If you think you can, or you think you can't, you're probably right.

Mark Twain

- **Address your fear**—Exaggerating the worst thing that can happen will demonstrate the absurdity of your fear. For example, fear of rejection could be stopping you from sending an abstract of a paper for that presentation at a conference that was one of your career goals. Imagine the worst things that could happen if it were rejected: my boss would judge me as incompetent, then I wouldn't get a raise, then I would quit my job, then I couldn't find another job, and then I would starve to death. By the time you complete this cycle, you'll no doubt realize how ridiculous it is that you are letting fear keep you from starting the task. In the words of Mike Ditka, "Success isn't permanent, and failure isn't fatal."

- **Look at the price of procrastination**—If you're entering an intersection in your car, the price of procrastinating about which way to turn will become very clear when the traffic light changes color. You're so aware of the price in this circumstance that you don't procrastinate. Looking at the less obvious price of procrastination in other circumstances (missed opportunities, for example) can also be very useful.

It's not what you are, it's what you don't become that hurts.

Oscar Levant

- **Look at the benefits of completing the task**—These fall into two categories: the pluses that you'll enjoy and the minuses that will stop. Getting that abstract submitted will be a positive step toward your goals and will end the tension you felt about meeting the deadline. Learning to swim will allow the fun of snorkeling on the trip to Hawaii and will end uncomfortable feelings around water.
- **Reward yourself**—One engineer we know found himself regularly procrastinating over his monthly report. He finally adopted the "B and R research" method. When he completed his report, he rewarded himself by a trip to a nearby Baskin and Robbins for ice cream. As he handed his report to his secretary, he said he was leaving for a meeting at B and R research and would be back in an hour. Promise yourself a weekend skiing or golfing when that paper is submitted.
- **Eliminate distractions**—In addition to clearing your desk, try moving your in-box out of sight. It is a common tendency to look at your mail as soon as it arrives. This breaks your concentration. Also there may be something in the mail that would be easier or more fun to do, giving you another reason to procrastinate. Put your out-box on your desk to remind you to get things done and off your desk.

Procrastination is opportunity's natural assassin.

Victor Kiam

DELEGATION

The secret of success is not in doing your own work, but in recognizing the right (person) to do it.

Andrew Carnegie

The reasons to delegate work generate little disagreement: it extends the results beyond what one person can do, releases time for other tasks, develops the skills of others, puts decisions in the hands of those who are closest to the action where facts and expertise are available, and speeds up decisions. Even though these benefits are clear, delegation is a poorly practiced skill. We hear the following reasons most often:

- I can do it better (or faster) than anyone else.
- I just didn't think of it.
- My staff is already too busy, and I protect them by doing it myself.
- I'm too busy to delegate.
- I want the respect of my staff that I get from doing it myself.
- I don't want to look weak or incompetent.
- It makes me uncomfortable.
- I don't have time to examine the problem to see what parts could be delegated.
- I will have to do most of the task before I could explain it to someone else.

Never tell people how to do things. Tell them what to do and they will surprise you with their ingenuity.

George S. Patton

It may be true that you *can* do the task better or faster than anyone else, but, then again, you may be overestimating the uniqueness of your skills and underestimating those of others. Clearly, if you don't break out of this trap, there is a very real limit on what you can accomplish. You must train your replacement, or you'll never be free to move on. It also sends a signal that you have low expectations of your staff's abilities. Several studies have shown that performance by others improves as your expectations of them and of yourself rise. One of the most graphic illustrations is "Sweeney's Miracle," a managerial and educational self-fulfilling prophecy (described in the *Harvard Business Review*).

James Sweeney taught industrial management and psychiatry at Tulane University, and he also was responsible for the operation of the Biomedical Computer Center there. Sweeney believed that he could teach even a poorly educated man to be a capable computer operator. George Johnson, a former hospital porter, became the janitor at the computer center; he was chosen by Sweeney to prove his conviction. In the morning, George Johnson performed his janitorial duties, and in the afternoon Sweeney taught him about computers.

Johnson was learning a great deal about computers when someone at the university concluded that, to be a computer operator, one had to have a certain I.Q. score. Johnson was tested, and his I.Q. indicated that he would not be able to learn to type, much less operate a computer.

But Sweeney was not convinced. He threatened to quit unless Johnson was permitted to learn to program and operate the computer. Sweeney prevailed, and he is still running the computer center. Johnson is now in charge of the main computer room and is responsible for training new employees to program and operate the computer.

Recognize that feeling uncomfortable or uneasy about delegation is a natural reaction. You are putting your reputation in the hands of others. One of the most difficult parts of delegating is accepting and accommodating the differences of others. Others will: 1) probably do the work differently than you would have; and 2) possibly do it better. When you can accept these two aspects, you are on your way to delegating effectively.

I've had a lot of experience with people smarter than I am.

Former President Gerald Ford

A key step in delegating is being sure that you are delegating to the right person—someone with intelligence, aptitude, willingness to learn with your guidance, and knowledge of the task subject. Once you've picked the right person, here are some key steps to successful delegation:

- **Be very clear about the tasks being delegated**—Provide a written definition of the results you expect, when you expect it (including intermediate milestones if appropriate), what the budget is, and who is responsible for the task. Don't describe the steps you expect taken, but rather describe the results expected. Set realistic deadlines that all agree to. Explain the importance of the tasks.
- **Follow-up**—Effective follow-up is an essential ingredient in effective delegation. The methods vary, depending on the task. On one design project with a tight schedule, each task manager filed a weekly report showing what was accomplished on each drawing, how many hours had been spent compared to the hours budgeted, and what, if any, problems are anticipated. Regardless of the method, do follow-up while giving latitude to use imagination in accomplishing the work. Be available for questions. If help is needed, pitch in.
- **Delegate the authority**—Many people want to delegate responsibility but not the authority. It doesn't work. Those doing the task must control the factors determining their success.
- **Make sure the tools and assistance needed are available**—People also need to know where to get them.
- **Match the capabilities of the people to the task.**
- **Have high expectations**— Express your confidence. As discussed above, their performance and your expectations are directly related.
- **Reward success through monetary rewards and recognition**—Be generous with your praise for a job well done.
- **Ask staff to bring you solutions, not problems**—The phenomenon of "reverse delegation" can occur if you allow staff to bring you problems. You'll end up doing the work you delegated! If they bring you a problem, ask for their recommendation or for them to define alternatives.

In large organizations, we've seen two drastically different and unfortunately common categories of delegation: 1) those "doers" who don't delegate well because they can't give up hands-on conduct of their work; and 2) those who are eager to get into a "management" position, so they can delegate and never do any project work again. The rarer and more valuable species is the person who can do enough technical work to keep in touch with the realities of the workplace and marketplace, and effectively delegate enough of the work to be an effective manager.

Promoting a "doer" into a management position can be a double disaster—a good project person is lost and a poor manager is created. The widely known "Peter Principle" describes this phenomenon of rising to a level of incompetence. Unless the organization is very large, the total delegator is also a drain on the organization. As the level of management responsibilities increase, the amount of time "doing" decreases while the time planning and delegating increases. However, our experience is that there is a limit to the amount of tolerable decrease in doing. A zero "doing" level will soon lead to an obsolete manager or marketer because many technical fields are in a constant state of change.

By working faithfully eight hours a day you may eventually get to be a boss and work twelve hours a day.

Robert Frost

To know how much to delegate and how much to do is difficult. Some signs that your delegation is falling short:

- You're working longer hours than your staff.
- You're taking work home regularly.
- You're constantly rushing to meet deadlines.
- You're doing work assigned to your staff.
- You're regularly being interrupted by questions about work you've delegated.
- Your top-priority items remain undone.
- You're the only project manager you can identify for the next big project.
- You're experiencing low morale or poor initiative on your staff.
- You're experiencing a high turnover of "rising stars" in your organization.

The last point reinforces the folly of not delegating, presumably to protect your staff from too much work. When he was chairman of Avis, Robert Townsend emphasized the need to delegate "as many important matters as you can because that creates a climate in which people grow." Mackenzie reports that a study of 500 groups of managers by Ohio State University found that the leaders who rated good or excellent were the ones who made the greatest use of delegation. Leaders who scored poorly were ineffective delegators. Aggressive, rising stars will be driven away by a manager who refuses to let go of challenging work.

I not only use all the brains I have, but all I can borrow.

Woodrow Wilson

CALENDARS, DIARIES, AND ORGANIZERS

. . . make that recollection as durable as possible by putting it down in writing.

Benjamin Franklin

A concept that will be a most productive step in improving your effectiveness is *write it down.* The concept of writing all appointments, to-do items, and daily notes of phone calls and conversations in a single journal or diary can save you many hours per year. There are a variety of diary/calendar types available. A format we've found very useful is shown in Figure 5-3, using forms prepared by Daytimers, Allentown, Pennsylvania. In addition to the daily pages shown, there is a monthly calendar so that you can keep track of future commitments. A section in the back is used for phone numbers and addresses. The format is not critical, the concept is. Before adopting this concept, I (Gordon) took most notes in a notebook; put some notes on napkins, post-it notes, and scraps; put appointments in a calendar; kept a separate to-do list; and noted expenses randomly, usually search-ing through a stack of receipts trying to reconstruct expenses incurred several days before. I occasionally lost items. I was unsure of the dates of some notes or the relationship of notes to specific other events, and I lost track of what I did with some to-do items. I have saved many hours and much frustration by keeping all items in one diary and carrying the diary wherever I go. Supplemental right-hand pages (Figure 5-3) can be inserted into the loose leaf binder for days when notes exceed one page. Where it is critical to put notes in a project file, a copy of the page with the notes can be made and inserted in the project file.

Keys to effective use:

- **Keep the journal with you and use it as a continuous record of your conversations**—When in the office, leave it open on your desk to remind you of to-do items and appointments. It will also be readily available to make notes of telephone conversations and meetings or to add items to do as they occur. It will be easy to jot down ideas as they come to you. When traveling, note your flight information on the calendar pages so that your schedule is in front of you, keep it on hand to record messages received and expenses, and write notes of conversations and meetings.
- **Note commitments immediately**—As soon as a meeting or deadline becomes

TUESDAY
NOVEMBER 1990
13

46th Week
317th Day

APPOINTMENTS & SCHEDULED EVENTS

List meetings, appointments, deadlines here

7	
8	
9	
10	Bill Edwards, Portland
11	
12	Lunch, Jim Williams
1	
2	
3	Alaska Airlines, Flight 109 to Seattle (3:20)
4	
5	Spokane Proposal Due
6	
7	
8	
9	
10	

TO BE DONE TODAY (ACTION LIST)

Prioritized Daily Action List here

- Note status of each item here — is it an A, B, or C?

✓ = completed
→ = Forwarded to future date

- A Check status of brochure
- A Complete budget report
- A Check schedule for Tacoma meeting next week
- A Get cash for trip
- B Bill - time of meeting next week
- B Bill - advise of Dennis Howard interest in job

✓ = called back

PHONE CALLS

Record phone messages then when on the road

✓ Perry Baker 509-682-4417
✓ Dennis Howard 303-987-1122

EXPENSE & REIMBURSEMENT RECORD

Record daily expenses

Mileage 41 miles
Lunch, Jim Williams $17.25
Parking $9.50 (airport)

DIARY AND WORK RECORD

Use this space to record notes of conversations, meetings.

Bill Edwards - Bill will provide specs for review by Dec.15; I'm to provide drawing needs by Dec.1; next meeting will be on Dec. 20. Bill to send spec outline ASAP.

Dave Harris - agreed to meet in Nov. 20 at his office at 10 AM. Has comments on draft report. Concerned that cost estimates are high.

Perry Baker - new pumps started up this morning. Running great.

Jim Williams - RFP for new project scheduled to be out on Jan. 15. Main concerns are that project manager be based in Portland and be available full time. Pleased with the last project we did.

Dennis Howard - interested in going to work for us. 12 years experience in pipeline design. Suggested he send his resume to Bill in Denver office.

FIGURE 5-3 Sample daily planner.

firm, write it down at the appropriate time and place. A quick glance at any given day will then show you immediately what commitments you have.

- **Note future to-do items immediately**—For example, if you're traveling and think of a task that needs to be done upon your return to the office, immediately turn to the date of your return and note the item in the to-do column.
- **Prioritize your to-do items**—As discussed earlier, spend the first few minutes each day prioritizing your tasks for the day. As they're done, check them off. If they are delayed, move them forward to the next appropriate day. Add new tasks as they occur and prioritize them.
- **Plan the next day**—At the end of the day, transfer the things you didn't get done to the next or another day.
- **Use cross-references**—Let's say a telephone conversation on May 6 results in a meeting on May 14. When noting the meeting on your May 14 calendar, add a reference (such as: See May 6) so that you can quickly find and review your background notes relative to the meeting.
- **Include your personal and professional goals in your daily to-do lists and schedules**—As discussed earlier, it's important to develop your weekly and daily strategies based on your personal and professional ideal scenes for all aspects of your life. We've found it useful to record our ideal scenes in a section at the back of our journals so that we can frequently refer to them as we plan our daily strategies.
- **Note and keep your commitments to yourself**—If one of your goals is to exercise three times per week, note the times on your calendar and give them as much weight as if they were a commitment to your job or a client. You may be surprised to find your effectiveness, attitude, energy, and outlook improve.
- **Note follow-up dates** for delegated items on your calendar at the time you delegate the task.

HANDLING PAPERWORK

The person working within an organization may face a steady stream of incoming paper—some of significance (often a small proportion!), some of interest but not very significant, and the balance neither interesting or significant. Napoleon is reported to have a rule that none of his incoming mail was to be opened for a period of three weeks. His theory was that most of the problems it would have raised would have solved themselves in that time. If you're in a service industry, you can't take the luxury of solving the problem the way Napoleon did. You need to efficiently handle your paperwork.

Some items require action, while others do not. The steady inflow can become disruptive to your effectiveness. I recently called upon a director of public works in his office to discuss a project. The meeting occurred in early afternoon. As I sat down, he told me his goal for the day was to clear out his in-basket, which had overflowed into a large pile on his desk. Not only had he failed to make progress, he was moaning that the stack had grown even higher as the day progressed due to interruptions (see next section on interruptions) and continued inflow of paper. Here was a man out of touch with his goals and priorities. He was so overwhelmed by the tidal wave of paper engulfing him that merely shuffling the paper had become his top priority for the day.

One common theory to handling paperwork is to handle each piece of paper only once—pick it up, decide what to do with it, do it, and move on to the next item. This is a bit idealistic but use it to the maximum extent that you can. A very high percentage of items can be handled only once and sent to the file, sent to the library, delegated for action, trashed, or an appropriate response noted right on the paper. For example, if it is a letter or memo requiring a brief, quick response, write the response on the letter or memo and give it to a secretary for typing or, in the case of an internal memo, send the handwritten response.

To quicken your progress through a stack of mail, try these sorting ideas: if an item has less than first class postage, trash it immediately; put journals or reports that *require* reading into a "to read later" stack (good airline or evening reading material). What will be left after this sorting are those items requiring some action beyond an immediate quick response or disposal—usually a very small percentage of the incoming paper. Review these items and prioritize them on your to-do list. Clear all but the top-priority item from your desk, setting the rest and the in-box out of sight behind you.

Set aside one time in the day to sort your incoming paperwork to avoid it being disruptive. Training a secretary to do the sorting for you can be effective, especially when you're traveling. When you return, the important action items will be in one stack, the to-read items in another, and junk mail in another (consider giving him or her authority to trash less than first-class postage items to eliminate this stack).

Cluster your activities related to the paperwork so that similar activities are done at one time. For example, use one block of time to write responses to letters, use another for phone calls. Grouping similar activities will increase your efficiency.

CONTROLLING INTERRUPTIONS

When we ask people to list their biggest time wasters, "interruptions" is always high on the list. Interruptions include telephone calls, unexpected meetings, drop-in visits by coworkers, and personnel problems. When asked to keep track of interruptions, managers typically find about ten per hour—about half of which are self-generated. Of course, some interruptions (a client calling with a question) are an essential part of doing your job. We're focusing here on the peripheral, less productive interruptions. As already discussed, setting your priorities and focusing on the top-priority task will do much to reduce your self-generated interruptions. What about those generated by others?

All phone calls are obscene.

Karen Elizabeth Gordon

The telephone can be both a tremendous time saver and time waster. Many seem to find that socializing over the phone beats working. Try eavesdropping on telephone conversations the next time you walk around an airport during the middle of a business day. A high percentage of people are there on business travel and a high percentage of those on the phone at this time of day will be on business-related calls. What part of the conversations sound like serious business? What part sounds like socializing? This unscientific method may give you a feel for the amount of time wasted during typical business telephone calls.

Makenzie lists several reasons why we give the telephone power:

- **Curiosity**—Who is calling me? What will I miss if I don't answer it? Is it the person I've been playing telephone tag with for a week?
- **Fear of offending**—I can't cut short this phone call even though this caller is long-winded and off the subject, because he or she will be offended.
- **Legitimacy**—This must be important, otherwise they wouldn't call. It is probably more important than what I am doing right now.
- **Need to be needed**—I feel important when people call me for information.
- **Great excuse**—I really don't want to work on this report/budget/calculation/paperwork right now, so any phone call is a welcome diversion.
- **Socializing**—I like to spend a lot of time getting to know everyone, then we'll work together better.

There are several methods to make better use of your time on outgoing and incoming calls.

To streamline your outgoing calls:

- Outline the points you want to cover in advance.
- Be pleasant but direct. An opening statement such as "I know you are busy too. I have one quick question. . ." may save both of you time by focusing the conversation.
- Skip asking how his or her vacation was or how the family is doing. Such inquiries are an open invitation to socialize.
- Remember, your call is an interruption to the person on the other end who may be eager to return to work as well.
- Make your calls shortly before lunch or at the end of the work day when the other person may be eager to stick to the point.

• Set aside a specific time to make calls so that you aren't tempted to cause a self-generated interruption by making a call on impulse.

We've seen statistics that say there is only a one in five chance that you'll reach the person you're calling on the first try. Likewise, there is a one in five chance that you'll be available when your call is returned—hence the widely practiced sport of "telephone tag" (trading return-my-call messages). To decrease your time wasted in this sport, find out when the person you're trying to reach will be available. Call the person back rather than waiting for them to call. Your message may be a C priority to them! Also, leave word about the nature of your call so that the person will be ready when you reach them. If you do leave a message for a return call, indicate the time you'll be available, to short-cut the tag game.

As with calls you generate, set aside a specific time to return calls—but *do* return them. We've all experienced people who simply will not return calls—giving the impression that they are disinterested and rude (correct impressions) and too important to be bothered (no doubt an incorrect impression). Your reputation will be the better for returning your calls.

For three days after death, hair and fingernails continue to grow but phone calls taper off.

Johnny Carson

Incoming calls are difficult to control. One commonly proposed method is to receive calls only during limited periods during the day and have messages taken the rest of the time. We don't recommend this approach if you are in a service industry. For example, if you're in the consulting business, timely responsiveness to your clients is a key part of your service. However, there will be times when you truly should not be interrupted. When you're in a very time-critical task, such as finishing a proposal or report for a client to meet an imminent deadline, it is no more reasonable for the caller to expect you to be interrupted than it would be to expect a surgeon, to interrupt an operation. These occasions should be less frequent for you than for a surgeon, but when they occur, have messages taken.

At other times, take your calls as they occur, but be prepared to handle them efficiently. Maintain control by asking what you can do for them or if they have a question you can help with. Interrupt long-winded callers and perception check with them: "So let me check if I understand. The main issues are . . ." If you lose control, or have a time limit, say that you're sorry but you have to leave for a meeting or take a conference call. We've heard of one extreme method to end an unwanted call—hang up while you're talking; the other person will assume you were cut off because no one hangs up on themself! A more appropriate approach for an untimely call is to simply say that the caller has caught you at a bad time and you'd like to return the call at a specified time.

The drop-in visit by a coworker can be a time-consuming interruption. You can minimize the time by asking what you can do for them. Answer their request briefly if you can. If not, schedule a time when you can give them your full attention. Try standing up when they enter rather than sitting and inviting them to sit. Use body language to indicate you're in the middle of something by keeping your pen on the paper or your hands on the computer keyboard, or a hand on the phone. If it's a bad time, be candid. Tell them it's a bad time and you'll stop in to see them in 15 minutes (or at some specified time). By going to their office, you'll maintain more control over the length of the interruption, as well as finishing whatever you were doing before you see them. It is much easier to leave their office. Set a time limit at the start of the conversation to keep it focused. If certain staff members keep coming back with one question at a time, suggest that they collect a list of questions before coming to see you. Also, ask them to bring suggested solutions and ask them what they think should be done.

When faced with a task that simply can't tolerate an interruption, the most effective method is to stay at home or at another location to work on the task until it is finished. If there is equipment or materials that can't be moved to an isolated location, come in early—drastically early if you have a substantial amount to do. Staying late isn't as effective because it is much more likely that there will be those who will drop in to socialize before going home. We've found that those in an office at 5 A.M., for example, are serious about getting a task done.

TIME-SAVING IDEAS

Time is a great teacher, but unfortunately it kills all its pupils.

Hector Berlioz

We've had the benefit of hearing many good ideas. Some of these are from earlier discussions in this chapter, others stand on their own, and several are summarized by Alec Mackenzie in *The Time Trap*:

- Use your prime time to do the difficult or unpleasant task.
- Get to the point directly in conversations, letters, and memos.
- Focus on one task at a time.
- Do the tough part of the task first.
- Be on time.
- Take speed reading courses.
- Increase your energy level through diet, exercise, and rest.
- Know your energy cycle—protect your prime time.
- Positively visualize successful completion of the task.
- Learn to say no.
- Refuse to do the unimportant.

- Handle each paper only once (do it, delegate it, or dump it).
- Use your commute or travel time for small jobs, correspondence, reading, or listening to educational tapes.
- If you are stuck on a task, set it aside temporarily. Get a fresh perspective.
- Throw away bulk mail, unopened.
- Eat light lunches.
- Keep all notes and appointments in one calendar/diary.
- Keep your long-term goals in mind as you develop your daily strategies.
- Keep an up-to-date, prioritized to-do list with you at all times in your calendar/diary.
- Delegate everything you can effectively.
- Write replies to letters or memos right on the piece of paper.
- Keep the commitments you make for yourself, including regularly scheduled times to exercise or relax.
- Confront procrastination.
- Return phone calls just before lunch or just before the end of the day.

References
1. Waitley, Dennis. 1983. *Seeds of Greatness.* New York: Simon and Schuster.
2. LeKein, A. 1973. *How to Get Control of Your Time and Your Life.* Signet.
3. Mackenzie, R. Alec. 1972. *The Time Trap.* McGraw-Hill Book Company.
4. Gawain, S. 1978. *Creative Visualization.* New York: Bantam Books.
5. On human relations. *Harvard Business Review.* 1979.
6. Garfield, Charles. 1986. *Peak Performers—The New American Heros.* New York: Avon Books.
7. Kerzner, H. 1989. *Project Management. A System Approach to Planning, Scheduling and Controlling.* New York: Van Nostrand Reinhold.
8. Mayer, J.J. 1990. *If You Haven't Got The Time To Do It Right, When Will You Find The Time To Do It Over.* New York: Simon and Schuster.

6

Productive Meetings

EVERYBODY LOVES TO HATE MEETINGS

Meetings are indispensable when you don't want to do anything.

John K. Galbraith

When we ask people to list time robbing items, meetings always seem to be high on the list. Many seem to agree that a

. . . meeting provides a great chance for some people who like to hear their own voices talk and talk, while others draw crocodiles . . . It also prevents those who can think and make quick decisions from doing so.

Lin Yutang
The Pleasures of a Nonconformist

There is no doubt that poorly planned and conducted meetings are a substantial waste of time. Kerzner reported a survey of 300 project managers in a variety of industries and found that they typically spent eight hours per week in a meeting. That equates to two and a half months of each working year, or 9 years out of a 45-year career! We have certainly spent far too many hours in meetings that were poorly organized and pointless, or whose relevance to our attendance was a mystery. Such experiences must be widely held because we usually find more emotional charge on the expression of meetings as time-wasters than we find on most others. Yet meetings are perhaps the most important single stage for professional people because it is where you see and evaluate and are seen and evaluated. They can be a precious resource, not to be taken lightly.

I really looked for people who could manage other people in meetings. It separates the wheat from the chaff.

Verne Opp
Former U.S. Air Force Secretary

Because of the significance, we've devoted this entire chapter to improving meeting effectiveness.

WHY HAVE A MEETING?

With all the negative feelings about meetings, one wonders if there ever is a valid reason to have a meeting. There are some:

- **Interaction is necessary**—There are some issues that you cannot resolve unilaterally and can be handled most effectively in a face-to-face exchange, rather than in an exchange of memos or letters. Contract negotiations between two groups offers an example. A proposed contract generates a reaction and an alternate proposal. The response to the alternate may depend on an intervention between the two parties. Also, in negotiations, a response may depend on nonverbal clues as well as the verbal, and a face-to-face meeting can produce a different result than a written or telephone exchange. It is easier to convince someone in person. It is harder to say "no" when face-to-face.
- **A group solution is needed**—No one person may have the expertise to solve the problem, and group interaction is required to come up with a solution. For example, a production problem in an office may require input from professionals, technicians, draftspeople, and clerical staff. Of course, another benefit is that people are much more willing to accept a solution that they participated in developing than one imposed upon them.
- **Team Building**—Generating a team feeling may justify a meeting. It is possible to build a sense of togetherness, trust, and belonging, as well as build a commitment to the group goal. We've seen projects with demanding schedules where a weekly team meeting was a valuable asset. Most of the *information* could have been distributed in writing; however, the *group energy* developed by interpersonal support in the meeting could only result from the meeting.
- **Information**—In some circumstances, it may be beneficial to have a meeting for distributing and clarifying information. We've seen many informational meetings that should never have happened because the information could have been distributed in writing just as effectively. There may be a few occasions where it is desirable to have everyone receive the information at the same time or where the information may generate questions that need an immediate response. For example, when a firm is acquired, an informational meeting gives everyone the news at the same time and provides answers to questions and

concerns. This eliminates a potentially tremendous amount of lost time in rumor trading, gossip, and anxiety over unaddressed questions.

ALTERNATIVES TO MEETINGS

Time is money.

Benjamin Franklin

In considering alternatives to meetings, keep in mind the cost. Meetings can be very expensive. Ten managers in your company, each earning $35 per hour, averaging eight hours per week in meetings will cost you $150,000 per year, plus the cost of lost time associated with preparing for meetings, interrupting their work or travel time, and travel expenses. The true cost of time for any meeting is greater than the time of the meeting itself. Premeeting and postmeeting time is required. When a meeting produces poor results, indirect costs can be significant. Morale can be depressed. A poor staff meeting can set a poor tone for the entire work week. Impacts can continue to cost you money for weeks or months. Consider the number of meetings called by people throughout your company everyday. The cumulative costs can be enormous.

There are several alternatives to meetings to consider:

- **Put it in writing**—Information distribution or straight-forward requests for information can often be more effectively handled in writing.
- **Telephone**—A direct call or a conference call can be much less costly than a meeting. For the price of one airline ticket from either coast to Denver, you can pay for several hours of conference calls.
- **FAX/Phone**—The FAX machine is rapidly becoming ubiquitous. When combined with telephone conversation, it is possible to exchange, discuss, and modify documents nearly as rapidly as if all parties were sitting in the same room.
- **Video**—Suppose you want to disseminate pictorial and verbal information on a project to several branch offices and the desired impact can't be generated in writing. Rather than travel to each office to present the material in meetings, a videotape of the presentation could be shipped to each office where individuals or groups could review it as it fits their schedules. This avoids disrupting your schedule and theirs, avoids travel time and costs, and presents the message in exactly the same way to all offices. With satellite technology, video conferencing can also be an economical alternative to large group meetings.
- **One-on-one exchange**—When only two people are involved in resolving a problem, it clearly is more efficient to do it on a one-on-one basis rather than collect a group that ends up as spectators for the meeting. This seems obvious, but we've seen many group meetings where only two people were involved in

the decision process and where they didn't need others for input. By filling the meeting room, you can bet that some of the "spectators" will speak and the meeting time will expand to fill the allotted time. Once the spectators realize that their "input" has been totally irrelevant to the issue-resolution, which laid in the hands of only two people all along, they will feel cheated of their time and be resentful. Before assembling a group for a meeting, be sure that you are truly seeking input from more than one person. It may be possible to resolve an issue in ten minutes in a one-on-one exchange. How many group meetings have you attended that are over in ten minutes?

Consider the alternatives carefully before proceeding with a meeting, but, if you must, proceed using the guidance provided in this chapter.

AVOID THE PITFALLS

A committee is a cul-de-sac down which ideas are lured and then quietly strangled.

Sir Barnett Cocks

Meetings are not a good place to do detailed analysis, resolve personnel problems, attempt to convert ideas into words or drawings, organize data, or edit reports. Steer away from these purposes.

If you have a valid reason for a meeting, avoid the common pitfalls encountered that have ruined many a meeting:

- **Too many participants**—A common rule of thumb is that the probability of meeting success is inversely related to the number of participants. We've heard of one government official who said he'd found the perfect way to put an issue into permanent limbo—appoint a task force with 50 members. You can be assured nothing will happen.

Committee: A group which succeeds in getting something done only when it consists of three members, one of whom happens to be sick and the other absent.

H.W. Van Loon

- **Wrong participants**—Uninterested attendees will impede a meeting. The right mix of people is important. A later section of this chapter provides guidance on who should attend.
- **Unequal participation**—Later in this chapter, we'll talk about typical problem individuals found at meetings, which includes the *bigmouth species*. If there isn't a strong traffic cop to draw all attendees into equal participation, the meeting will fail due to intimidation of the weaker species.
- **Fear of attack**—It takes attention and effort to create an atmosphere of safety

for all meeting participants. If individuals feel they will be attacked personally rather than having their ideas discussed, the meeting will not be productive.

- **Inadequate preparation**—Successful meetings don't just happen. It takes more than setting a time, place, and topic. This chapter provides guidance on how to prepare an agenda and plan the meeting.
- **Lack of focus**—Without a clearly stated focus of the meeting, agreed upon by the participants, they will assuredly each charge off in their own direction once the meeting begins and there will be no progress. The same result will occur when there is not a clear definition of who is responsible for what at the meeting.

If Moses had been a committee, the Israelites would still be in Egypt.

J.B. Hughes

- **Topic and authority don't match**—A group meets to discuss an issue that they have no authority to resolve. Why bother?
- **Poor meeting environment**—Meeting room is too small for the number of attendees, or too large, too hot, too cold, and so on.
- **Wrong time**—Evening meetings compete with fatigue and personal plans, those immediately after meals may suffer from low energy levels.
- **The routine, prescheduled meeting**—Be suspicious of routine meetings, like the weekly staff meeting that occurs every Monday at 8:00 A.M., for example. It is almost assuredly a time waster, since the probability of a topic of significance appearing each Monday at 8:00 A.M. is infinitesimal. There are usually far more efficient ways of achieving the information exchange, which, at best, is the most redeeming feature of such meetings. Typically, such meetings are held too frequently, are poorly planned, include too many people, and last too long.

Let's now turn to the ingredients of a successful meeting.

IDENTIFY THE PURPOSE OF THE MEETING

He was wont to speak plain and to the purpose.

William Shakespeare

Have you ever left a meeting wondering why the meeting was held? The agenda was full of interesting topics, but no decisions were made. The information was found in routine written reports. The social interplay between attendees was pleasant, but you didn't need a meeting for that. There are an amazing number of meetings, including those at high levels in major corporations, where there is no clearly delineated purpose in advance.

It's a bit like going to a movie without knowing whether it's a comedy, romance, thriller, tragedy, or western. You could show up looking for a laugh and end up sitting through a tear-jerker. Not very fulfilling. The meeting will be much more productive and rewarding if all know the purpose in advance. For one thing, you won't end up spending part of the meeting time resolving why you are there.

You need more than a topic, place, and time for a productive meeting. Why are you meeting? You may want:

- **To give or exchange information**—Typically, meetings are overused for this purpose. There are usually more effective ways to distribute or gather information. People's ability to comprehend written information is four times greater than to comprehend oral presentations.
- **To create or develop ideas**—Meetings can be the best way to brainstorm. The most difficult challenge is to recognize that brainstorming meetings need to be structured differently than most meetings. They require openness and freedom so that all ideas are presented without any judgments (see Chapter 9 for more discussion). It's often best to separate brainstorming meetings from any other purpose. Delay screening of ideas and decisions to a separate time to avoid restricting the flow of ideas.
- **To make decisions**—Meetings can be useful for this purpose if the decision requires more than one person and when increasing commitment to a decision is vital. It's critical that all the preparatory work needed to make the decision has been done and that all have had access to this information before the meeting.
- **To delegate work or authority**—Meetings can allow for effective clarification of assignments and responsibilities and to gain the commitment of the groups to the assignments. Be sure that it isn't a substitute for a manager's inability to make decisions.
- **To share work**—Meetings can be used to jointly do work but groups of any size are an ineffective way to do work. Look for someone who is trying to avoid work if these meetings occur often.
- **To persuade or involve**—It is easier to persuade or increase someone's involvement with a face-to-face meeting. Thorough preparation is essential.
- **To consult**—When a problem can benefit from input from people with several different perspectives or expertise, a meeting can be effective. The keys are to clearly define the problem and keep clear separation between the processes of consulting and making decisions.

The approach to each of these purposes is different in terms of who should attend, how many should attend, and how the meeting is conducted. If it is strictly an informational meeting, you could have ten or ten thousand attend, and the meeting conduct would be the same; therefore, other then the cost of the meeting, it doesn't matter who attends.

A problem-solving meeting does not necessarily require a decision be made. Often, it is better to separate the two functions. The meeting may simply be to brainstorm alternative solutions to the problem and review their pros and cons. There may well be several who should participate in brainstorming, but who would not be involved in decision making. It is important that those attending agree that the subject is a problem and that they want to do something about it. If you're the only one that considers the company logo to be a problem, there isn't much point in having a meeting about the logo.

If the purpose of the meeting is to make a decision, let everyone know in advance how the decision will be made and who will make it, and get the agreement of the attendees. If only 3 of the 12 attendees are going to be involved in the decision, make it clear that these 3 want input from the others, but will make the decision themselves. It is pointless to hold a meeting to make a decision that the attendees have no authority to implement.

If a meeting is called to discuss long-term plans and goals, it is usually prudent to include many people to gain their commitment. If the planning is dealing with a short-term logistical project problem, it is best to keep the number of attendees to the minimum, to speed the process.

PREPARING THE AGENDA

The typical corporate meeting in America starts at 11 A.M., lasts an hour and a half, is attended by nine people, and follows a written agenda less than half the time.

Conclusion of a University of
Southern California Survey

A properly prepared agenda is as valuable as the road map for your automobile trip from New York to Los Angeles. Both will show you where you're going and how you're going to get there. The meeting leader is usually responsible for the agenda. However, people will better support what they create. When possible, solicit attendee input and comment on the agenda prior to its finalization. The agenda serves several purposes:

- A guide to prepare for the meeting.
- Tells participants in advance what is to be considered and what is expected of them.
- Provides a structure to maintain order and control.
- Focuses and facilitates discussion and keeps the meeting moving along.
- Provides a standard to judge the success of the meeting.

Agendas are so important to a productive meeting that we suggest you decline to attend a meeting for which you haven't received an adequate agenda far enough in advance of the meeting. You need to have time to raise questions, receive clarification, or suggest changes. The following are the essential elements of an agenda.

- **Topic of the meeting**—A descriptive title is appropriate, such as "Resolve Drawing Format."
- **Purpose of the meeting**—Keep it simple. Is it decision making, problem solving, information sharing, planning, or something else (as discussed in the preceding section)? If it's decision making, identify who makes the decision and how. *Clarity on purpose is a prerequisite to a successful meeting.*
- **Who called the meeting**—It's important to identify who called the meeting so that invitees know who to contact for information, to suggest changes to the agenda, or to let them know if they will be attending or not.
- **Who should attend**—In deciding on attendees, consider the following questions:
  - Who has relevant expertise?
  - Are we duplicating expertise or points of view by others present?
  - Who needs to participate in the decision?
  - Does the timing and individual workload assure full participation, or will the attendee be there in body only?
  - Who is crucial to implementing the results?
  - Who represents a cross section of opinions or views (a homogenous group preordains the outcome)?
  - What are the relationships of the attendees? Subordinates could be intimidated by a boss, limiting their input.

  Group dynamics become very complex if there are more than 15 in attendance, when problem solving or wide-spread participation is desired. As groups get larger, members feel less responsible for making it work and people start acting as members of subgroups rather than as individuals.

  It can take a professional facilitator to get productive problem solving or decision results from groups larger than 15. Groups smaller than 15 are much better for problem solving and decision making because they allow everyone to participate, spontaneity and informality can be maintained, and group synergy seems to develop more readily.

  For informational meetings, groups over 15 are easy to deal with.

- **Location, time**—The date, starting time, and location of the meeting is basic information. If you suspect that attendees may not be familiar with the location, include a map or directions. Include special parking instructions, if needed. The right location can be an important ingredient of a successful meeting. The room size should reflect the number of attendees. An overcrowded room can become hot and stuffy, and disrupt progress. On the other hand, a small group in a huge room can be intimidated by empty space. Pick a room where you can control the temperature. Most experts recommend a meeting room temperature of about 68 degrees. Make sure there isn't remodeling work or a noisy party scheduled in the room next door. If you're going to be using slides, be sure that you can control the light level.

The time should be selected to best suit the schedules of the attendees. If it's a project meeting that requires the attendance of staff from several public agencies, Wednesday evening would be a poor choice of a time! However, if you want to have a project meeting for neighborhood input on a project design, Wednesday evening may be an excellent choice. Several management books suggest picking an odd time (like 10:03 A.M. rather than 10:00 A.M.) for starting a meeting, on the theory that people will be more likely to be on time because the starting time appears more precise. Our experience is that changing the starting time to an odd time is much less effective than consistently starting the meeting on time (as discussed later).

- **Ending time**—Set a realistic ending time and stick to it. A fixed time constraint will do wonders for focusing the discussions. In determining the overall time for the meeting, consider scheduling five- to ten-minute breaks if the meeting time is going to exceed two hours. There is a finite limit to attention span and physical needs.

- **Agenda items**—If possible, keep the agenda items within the same subject area, to minimize the number of essential attendees. Consider two separate meetings if topics are very diverse. Also, limit the number of items because meetings fail in proportion to the number of items undertaken. There should be a brief (two- to three-sentence) narrative description of each agenda item. The description of each agenda item should be specific and should include a realistic assignment of the maximum time to spend on the item as well as identifying who is responsible for leading the discussion or action on the item. If there are backup materials associated with the item, they should be noted and attached to the agenda. Each item should be noted as to whether it is an action item, information item, or discussion item. Be sure that the sum of all the individual agenda items are less than the total time allotted for the meeting! Often, the first attempt at having an agenda including all the potential items will result in a need for more time than is available. Resist the temptation to cut the time for each item because it is rare that the needed time will be less than estimated. Instead, defer the lower priority items to a separate meeting. The order of the agenda items usually influences the attention and energy devoted to each, and you can use the order to suit your preference for the meeting outcome. Items that you want to receive the most attention and debate should appear early in the agenda, when the energy levels and attention are greatest. Items that must be resolved should be near the beginning of the agenda. It is prudent to start the agenda with an item likely to be completed successfully because its success will set a positive expectation for the entire meeting. If you want to minimize the attention paid to an item, put it last on the agenda. End the meeting with an item that will make people feel like they accomplished something. If you can't, close the meeting by summarizing the positive aspects.

An example agenda is made up of items shown in Figure 6-1.

• **Circulate the agenda**—By circulating the agenda in advance of the meeting, you'll increase the chances that attendees will be prepared and that any suggestions for adjustments to the agenda will be received in advance. How much in advance is enough? The amount of time is dependent on the nature of the agenda. If there is a substantial amount of backup material, allow ample time for people to read it, recognizing that their lives involve more than just your meeting. Bring extra copies of the agenda to the meeting for those who forget theirs.

Monthly Manpower Committee Meeting
Date: November 17
Location: Conference Room B
Chairman: Jim Wilson
Starting Time: 7:00 P.M.

1   *Minutes of Last Meeting*. ACTION ITEM—Chairman Wilson   (5 minutes)
       Approval of minutes of September 6, 1991.
       Minutes attached.

2. *Budget for 1992*. ACTION ITEM—Chairman Wilson   (30 minutes)
       The revised budget incorporating comments from past
       meetings is attached. Following discussion, a budget
       will be adopted by a majority vote of those present.

3. *Benefits Committee Report*. INFORMATION ITEM—Bill Phillips   (20 minutes)
       The committee will present the results of its
       current benefits compared to those commonly found.

4. *Long-Range Planning*. DISCUSSION ITEM—Tom Williams   (40 minutes)
       The purpose of this discussion is to identify the critical
    issues to be addressed in the long-range planning effort
    scheduled to begin next month.

Adjournment: 8:35 P.M.

FIGURE 6-1. Sample Agenda.

CONDUCTING THE MEETING

Key Roles in the Meeting

There are four major functions that must be carried out in every meeting:

- Chair the meeting.
- Facilitate the process of the meeting.
- Record the result.
- Active participation.

In some cases, the chairperson and facilitator roles are combined into one individual. Some argue that the two roles should always be separate so that the individual facilitating the process of the meeting is not also an active participant.

The chairperson, in addition to being an active participant, sets the agenda, can argue a point of view, and is the spokesperson for the group. Facilitating the meeting involves keeping the group focused on the task at hand by suggesting procedures, protecting group members from attack, and acting as a traffic cop to make sure everyone has a chance to speak. The recorder is, ideally, not an active participant and is there solely to record enough of the meeting so that the ideas and results can be accurately recalled. Of course, the group members are the active participants who determine the course of the meeting.

In small or less formal meetings, the facilitating role is often combined with the chair role, and the recorder is one of the group members. It is a challenge for an active participant to slip in and out of the neutrality of an ideal facilitator or recorder, but practicality often dictates the combined roles. Failing to be neutral in the facilitator or recorder functions is a common problem in meetings, especially when the roles are filled by active participants. It is very difficult to have a vested interest in the subject of the meeting and remain neutral as a facilitator or recorder, hence the risk.

Preparing to Chair the Meeting

Meetings tend to fail in inverse proportion to preparation time and in direct proportion to meeting time.

George David Kieffer

The success of a meeting depends on the open and spontaneous interaction of participants. The meeting leader can stimulate or squelch this interaction by his or her actions. Prior to the meeting, the chances of success can be improved if you:

- Know the agenda, the subject matter, and any related history of past meetings or actions.
- Have a clear picture of the objectives of the meeting.

- Think through the probable course of the discussions and how it could affect the conduct of the meeting. Visualize a successful meeting and how it would proceed.
- Appraise the interests, biases, and personalities of the attendees and how they may interact. Ask yourself: Why are they at the meeting? What do they want to accomplish? What are their roles in relation to you and others? What biases or prejudices do they have relative to the meeting issue? What do you and they have in common? What can you and they share, to put yourselves on the same team?
- Consider each attendee as an individual and how to best relate to each.
- Know and use the language of the group.
- Check the meeting facilities in advance.
- Arrive early to make a last-minute check of the meeting room and to mingle with early arrivals, to establish rapport.
- Review visual aids in advance—Are they clear? Do they serve a purpose?

It is impossible to overprepare. Ask yourself before the meeting what is expected in the way of preparation and then exceed the expectation. Your credibility and the meeting success will both benefit.

Problem Types

Having served on various committees, I have drawn up a list of rules: Never arrive on time; this stamps you as a beginner. Don't say anything until the meeting is half over; this stamps you as being wise. Be as vague as possible; this avoids irritating others. When in doubt, suggest that a subcommittee be appointed. Be first to move for adjournment; this will make you popular; it's what everyone is waiting for.

Harry Chapman

If you've attended very many meetings, you no doubt have observed several types of individuals that can disrupt meeting productivity. Inspired by Doyle and Straus, Table 6-1 summarizes characteristics and cures for ten of the most commonly encountered individual types in our experience. The communication skills described in Chapter 2 offers some basic tools in dealing with problem types. Practicing the basic communication skills as a daily routine is excellent preparation for facilitating meetings as well.

For example, perception checking can be a very useful tool in leading a meeting. Don't assume that you understand a group member's point. Feed your perception back to them. It may stimulate some illuminating points from the members.

By setting up guidelines at the start of the meeting, you can reduce the chance that meeting robbers will control the meeting. It's important to spell any guidelines out and get all to agree to follow them. We've found that guidelines like these to work:

TABLE 6-1 Meeting Robbers

	Action (A) Results (R)	Cures
Tardy Tom	A Enters late and wants to be brought up to date. R Disrupts meeting.	• Ask after the meeting, why late? What would it take for them to be on time? • Check ahead to schedule meeting time. • Make them recorder of the next meeting. • Consistently start all meetings on time.
Bailout Bob	A Leaves before meeting is over. R Drains energy from group.	• Ask after meeting, why leaving early? Are meetings too long? • At start of meeting, ask if all can stay until the end. Commitment will help. • Ask them to summarize their position or part before they leave.
Pete and Repeat	A Makes same point over and over. R Wastes time of group.	• Use the group notes to show that the point has been noted. • Give the person two to three minutes to say what he/she has to say and let them know it is time to move on.
Doubting Thomas	A Negative on everything. R Reduces group creativity.	• Prohibit evaluation of ideas for an agreed upon portion of the meeting. • Restate the objections raised and ask the group to comment. • Bring the optimists into the discussion.
Sidetalker	A Talks to neighbor during meeting. R Breaks group concentration.	• Walk up close to talkers. • Ask group to maintain a single focus. • Get group agreement at start of meeting, "no sidetalking." • Make sure chronic sidetalkers don't sit together. • Ask one of them a question.
Big Mouth	A Talks too much and too loud. R Intimidates others, reduces participation.	• Solicit comments from others by calling on them. • Appoint the loudmouth as secretary/recorder to take notes.
Enemy	A Makes personal attacks on chairman or other group members, or asks very hostile questions. R Creates tension in the group and can cause sidetalking.	• Move away from them to decrease confrontive air. • Let them vent feelings. • Ask them for their alternatives, get them involved in the process.

TABLE 6-1 Meeting Robbers (continued)

	Action (A) Results (R)		Cures
Busybody	A	In and out of meeting.	• Adjourn meeting until they can attend.
	R	Wastes time of group.	• Confront before the next meeting, explain disruptive effects of behavior, get commitment to stay in meeting.
			• Schedule meeting outside of office or outside of normal hours.
Interrupter	A	Cuts others off or jumps into pauses that are part of someone's thought process.	• Ask them to hold on until the other person is through and confirm that you'll get back to them.
	R	Causes hostility and/or reduces participation by quieter members.	• Invite person interrupted to comment.
			• Appoint the interrupter as the recorder to take notes.
Dropout	A	Refuses to participate, eyes glazed over, doodles, reads a book.	• Ask them questions during the meeting, preferably directly related to their area of interest.
	R	Depresses group energy.	• Assign them part of the agenda.
			• Talk to them at a break; ask why they aren't participating—maybe they shouldn't be there.

- No interruptions once the meeting begins. No one leaves early.
- When you want to speak, raise your hand. Wait to be called on.
- Stick to the time limit for presentations. You will be cut off when time has expired.
- For any decision items, agree on the method of decision—majority vote, chairperson decides, and so forth.

Starting and Ending on Time

Consistently start your meetings on time. The starting time is the first test of your control of the meeting. Starting on time will set the expectation of the group of your ability to perform. If you delay a few minutes to start one meeting, you'll find that people will begin arriving at the next meeting according to their best guess as to when it will start. Start with whoever is present at the appointed time. You'll be pleased to find an increasing number of on-time arrivals at your meetings as your practice becomes consistent.

By consistently ending your meetings on time, people will realize that they will have to focus on the agenda items. Nonproductive chatter will be discouraged.

Keeping the Meeting on Track and Moving

There is a tendency for every meeting to drift toward collective incompetence.

George David Keefer

Many psychologists have noted that groups of individuals are far more likely to make bad decisions than the individuals would alone. This tendency is not inevitable, but it will always be present to be dealt with. It is also why it takes far more effort to have a successful meeting than you might anticipate.

Among the factors that affect group results:

- **Miscommunication**—Each individual has his or her own communication ability and style. Some miscommunication will occur. Frequent perception checks are needed to avoid drifting into communication oblivion. Protect the quiet members. Solicit their input by directly asking for it.

- **Outside pressures and personal agendas**—Every attendee will have some outside concern relating to the meeting. You can't remove these, but you can analyze what these pressures are so that you deal with them. For example, is Al worried about losing his job to Bill? Such insights will help you analyze what's behind a criticism or suggestion during a meeting.

- **Basic insecurities and needs**—Every attendee will be concerned about economic well being, a sense of belonging, the need for recognition, and control of his or her life. The desire to belong and be liked may cause some to go along with a group decision, no matter how poor the decision appears to them. By recognizing these needs, you can address them and support the purpose of the meeting.

- **Distractions**—Any number of things can cause group members to be distracted. It could be that a key word will cause one member's mind to be side tracked. Others may be distracted by a poorly fitting chair. Assume that some of your group is lost at all times. Do not assume that the group mind is functioning as a whole. Perception checking and feeding back results can help align the group mind.

- **Avoidance**—Have you noticed that the time spent in meetings on an item seems to be inversely related to its importance or complexity? In groups, complex matters tend to be ignored and simple ones belabored because people enjoy talking about what they know best rather than what they don't know. As a result, groups tend to look for the comfortable issues, avoid the tough ones, and drag the meeting toward triviality. Recognize this tendency in conducting your meetings. Focusing the group's energy and attention on the important issues will require your substantial effort to keep them from drifting to the trivial.

The chairman . . . should set an example of courtesy, and should never forget that to control others, it is necessary to control one's self.

Roberts Rules of Order (1876)

To effectively lead the meeting, you'll need to combine the skills of coach, referee, and quarterback. You must embody a commitment to each member of the group and to a common objective. Your primary concerns should be the welfare of the group, the integrity of the meeting process, and achieving the group objective.

Managing a meeting is simply managing and communicating with people. Apply the basic skills discussed in Chapter 2. Empathize with participants, establish rapport, and check perceptions. Be responsive to the emotional energies you sense are present. Use courtesy, common sense, and patience. Be neutral, even when someone is blowing off steam.

Be alert for these common communication problems:

- Members using jargon or words with unclear or ambiguous meanings. Clarify the language by checking your perception of the meaning.
- Excessive generalization. Ask that the point be related to the specifics at issue.
- Members assuming that other members are thinking the same thing. Ask other members if they are in agreement with the point of view.
- Members missing the context in which a statement is made. This will be apparent by the responses of other members. Present your understanding of the context and ask if it is what was intended.

The group must keep a common, clear focus. If they don't, each member will be wandering off in his or her own direction. The meeting will soon look like a car where every passenger has a steering wheel. It is part of the chairperson function to be sure that the focus is maintained. Keeping the group on one item at a time is a key. Identify the agenda item being discussed and don't allow the discussion to wander onto other topics. Identify how the agenda item is to be handled before getting into the meat of the issue.

It's important to be sensitive to the pace of the group. If they appear to be losing momentum, take a break. If they appear to be rushing through important issues too rapidly, remind them that there is time for careful consideration, so as to avoid later regrets.

Essential to the meeting's success is each person's opportunity for equal, safe participation. If someone is dominating the conversation, thank them for their contribution and ask someone else (one of the quiet ones) what they think on the issue. People are not going to participate if they have to exert all their energy just trying to get a word in or if they feel they are going to be attacked personally for speaking. If the facilitator is concerned that everyone is heard, the group members will be much more likely to participate. By being positive and encouraging, you can provide a feeling of safety.

Many times we've seen someone pause to think only to have someone else jump into the few seconds of silence to begin speaking. A good response is to ask the interrupter to please hold his thought until the first person is finished and you'll get back to them next. You can solicit member participation by firing questions back to them—when one member raises a question, acknowledge that it's a good question and ask others in the group to answer it. The group members should do 95 percent of the talking in a meeting.

If someone comes in late, don't summarize the meeting for them or comment on their tardiness. You'll just be adding to the distraction. If you know someone will be arriving late, let everyone know at the start of the meeting. It will reduce the later distraction and let everyone know the late arrival isn't a criticism of the meeting or its attendees.

Focus on uniting the group as the meeting progresses. As chair, the group will tend to focus on you. What you say or do will become a reflection of the group itself; so, you should project what you believe the group should think of itself. Nearly all of what you say should be reflective of the group rather than of you as an individual. Watch for emotional buildups. If you see signs of anger or frustration, address them as quickly as possible. One approach is to simply say, "Tom, you seem upset. What are you feeling?" Give Tom the chance to vent his feelings and ask him some follow-up questions until you sense he has released his emotions. Until the emotions are released, it is counter-productive to present a different point of view or attempt to deal on the mental level.

Offer positive feedback to members at every opportunity by noting their contributions and by being generous with compliments. When a particular task or item appears complete, summarize the conclusion and make sure everyone agrees before going on to the next item. Move on to keep the momentum going.

To get a good decision out of a group, don't argue your own point of view. Feel free to present your view, but listen to and consider the views of others. Don't assume that, because there are differing views, someone must win and someone must lose. Look for the best alternatives available. Look for consensus options — ones that everyone can accept without feeling that they have lost anything crucial. Developing a consensus requires that all team members have the chance to voice their points of view, that a solution is found that is acceptable to all of the team members (even if it isn't everyone's first choice), and that enough time is allowed for team members to discuss the problem so that they buy into the solution. Juries are living proof that consensus is achievable. If you're considering several alternatives and are trying to reach consensus, take a straw vote on how many of the group couldn't accept each alternative. Focus on the one with the least opposition. Find out the reasons held by group members opposing the alternative. See if the group can develop modifications to address their reasons, turning the opposition into positive alternatives. Although it is a trial and error process, solutions leading to consensus can often be developed. If they are not, at least the decision maker knows where everyone stands and can make a more informed decision.

At the end of the meeting, check your perception by summarizing the results. Check for any second thoughts, questions, or suggestions. Establish who is responsible for any action items and when. If appropriate, determine when and where the next meeting will be held. Ask for any suggestions on how to improve the next meeting. Remember, it is the last statements of a meeting that a group tends to remember, so end with positive feelings, even if the meeting was less than a success. By the end of a meeting, people will react and remember your tone and energy more than the substance of your comments. Program the group for success in the next meeting!

KEEPING NOTES

The recorder plays an extremely important role in the meeting. His or her notes serve as the collective memory and can be a very valuable working tool for the conduct of the meeting itself. You may be asking how notes can be a valuable tool if your frame of reference is limited to the classical, typed meeting minutes that are distributed several days after the meeting is over. We have found, as have several others, that group notes written on large sheets of paper with bold, felt-tip pens as the meeting progresses are invaluable assets to a productive meeting. These notes provide a focal point for the group.

Have the group sit in a semi-circle facing an easel, with the pad of large paper being used for the notes to focus their energy on the subject. As one page is filled, rip it off and use masking tape to attach it to the wall so that the full running record of the meeting is visible for all to see throughout the meeting. If the meeting pace is fast, you may want an assistant to place the sheets on the wall. This will keep everyone, even the latecomers, abreast of what's been discussed and how the group got to its current position. It lets each group member see that his or her idea has been heard and recorded. It's a valuable tool to cut off someone who's repeating the same point over and over. You can point out the written record of the point and ask that the group move on. It provides an instant check of the accuracy of the notes because they are being written in full view of the group. It also provides continuity. If there is a break of a few hours or a few days, the notes from the last meeting can be placed on the wall to refresh everyone's memory. They also provide an excellent basis for the formal minutes of the meeting.

The note-taker (recorder) needs to know the terminology and subject matter or will have difficulty keeping up. When taking such notes, don't edit or paraphrase. If you're not sure of what was said, do a perception check before writing down the notes. Don't record the name of the speaker because you want the idea to be transferred to group ownership rather than being invested in personalities. Don't be bashful about asking the group to slow down if you're falling behind. If a decision is made, circle or underline it. Try using different colored pens to divide topics or ideas. Be sure that action steps are clear and agreed upon by all. Be sure to number the sheets so that the order can be maintained when they are taken off the wall.

After the meeting is over, the large sheets should be used to develop the minutes of the meeting or a memorandum describing the results. Just copying the group notes probably won't do because they are often abbreviated and will require some commentary to be comprehensible to someone who wasn't at the meeting. The memo should describe how the group defined the problem, the alternatives considered, how alternatives were evaluated and what criteria were used, and the decisions made. It's best to translate the group notes into minutes as quickly as possible, while the memories of the meeting are fresh. Timely preparation is especially important when action items have been assigned.

BEING A PRODUCTIVE GROUP MEMBER

Any time you're in a meeting, you're sending a message about who you are, what your abilities are, and what league you belong in.

Harold M. Williams

If you've been in many meetings, you no doubt have noticed that some group members are more productive than others. The recorder and facilitator have more visible roles, but you as a group member carry an equal responsibility for a productive meeting. First, know why you are going to attend and what you want to accomplish. If the agenda is unclear, ask the chairperson prior to the meeting what his or her objective is. Determine if your attendance is necessary. When you have made the decision to attend, be fully there and focused on the meeting. Give it all you've got. If the meeting requires materials to be reviewed in advance (i.e., discuss a draft report), review the material carefully before the meeting. If the advance materials are not received on schedule, ask that the meeting be delayed until the materials can be reviewed. Sticking to a predetermined meeting schedule without the opportunity to prepare insures failure. If a subject is raised during the meeting that requires preparation, state that discussion of the subject needs to be delayed until the preparation is done. If an item not on the agenda is raised, ask that it be scheduled for a later meeting because the right people may not be there.

It's important that you speak up if people are being cut off, if the chairperson is trying to manipulate the outcome, or if the group notes being taken are inaccurate. Practice active listening (see Chapter 2). Remember how irritating it is when others talk among themselves while you speak up at a meeting. Give the speaker your full attention. Treat others with respect. Be the kind of participant you want others to be. If you disagree on a point, address the point and not the person. Acknowledge the positive aspect of the point before raising your concerns. This will avoid the fear of personal attack that inhibits full participation. Conversely, when one of your ideas is being criticized, don't take it as a personal criticism—separate your idea from you as a person. Accept emotions directed at you without reacting emotionally. Look upon the criticism as a opportunity to learn how to improve your idea. If you feel yourself starting to lose this perspective, take a deep breath, relax, and invoke a calm, open mind for yourself. Praise other members and the group. Congratulate the group on progress made to create a positive atmosphere.

Authority is 20 percent given and 80 percent taken.

Peter Ueberroth

In an extreme case, you may find yourself in a position of being in a meeting where the facilitator or chairperson is doing an abysmal job. By using your basic communication skills and your facilitator skills, you can turn an unproductive meeting around from your seat as a participant. For example, you can speak up, bringing those wandering back to the subject; you can ask other group members for their ideas, to increase participation; you can cut off repeaters by pointing out that their ideas are recorded in the group notes, and so on.

If, as a group member, you are called upon for a presentation as part of the meeting agenda, be prepared. Review Chapter 2 on presentations and apply these techniques. Trying to wing it is discourteous to the group and decreases the chances of success. When your part of the meeting is over, ask if you are still needed.

WAS THE MEETING SUCCESSFUL?

Meetings are like panda matings. The expectations are always high, but the results are usually disappointing.

Joe Taylor Ford

The key to knowing the answer is to establish a standard *before* the meeting by which you can measure success or failure. Two of the key questions to ask after the meeting to determine success are:

- Was the standard met? Were the desired results achieved? (Check the purpose stated on the agenda—was it realized?)
- Did the members of the group act synergistically?
  If the answers are yes, the bottom line is "a success." If either answer is no, then a careful analysis is needed, reviewing the guidance in this chapter:
- If a decision was to be made, was the decision maker there?
- Did everyone understand and agree with the decision-making process?
- Were the right people there?
- Was the group too large or too small?
- Was the meeting space suitable?
- Was the agenda sent out in advance? Were the agenda items appropriate in nature and time? Did the agenda clearly state the purpose of the meeting?

- Did the group agree upon the problem in need of resolution (if the meeting purpose was problem solving)?
- Did the group focus on one item at a time?
- Were group members attentive to those speaking?
- Were the presenters adequately prepared?
- Did everyone have a chance to participate?
- Was the meeting conducted in an impartial, safe manner?
- Did the behavior of a few participants disrupt the meeting?
- Were accurate group notes taken, using the techniques described in this chapter?
- Were action steps clearly defined and responsibility assigned?
- Were minutes quickly and accurately prepared?

Answers to these questions will provide some insight into where the process may have gone off the track. It can be useful to videotape your meetings to analyze what is working and what isn't.

Good judgment comes from experience, and experience comes from bad judgment.

Barry LePatner

References

1. Doyle, M., and D. Straus. 1982. *How to Make Meetings Work.* Jove Books.
2. National Institute of Business Management. 1990. *Mastering Meetings.* New York: Berkley Books.
3. Kieffer, G.D. 1988. *The Strategy of Meetings* New York: Warner Books.

7

Project Planning, Scheduling, and Budgeting

FAILURE TO PLAN IS PLANNING TO FAIL

. . . it is not the practical workers but the idealists and planners that are difficult to find.

Sun Yat-sen

Many people do not like to plan projects. They prefer to do something "constructive," such as jumping into the project work, rather than "squander" their time planning. Sometimes, project plans are prepared merely to satisfy their boss, company policy, or their customer. In this case, planning is seen as a necessary evil. Much like insurance, resources spent on planning aren't thought of with much pleasure and aren't appreciated until disaster strikes. When a project plan is prepared merely to fulfill someone else's requirements, it is not followed and becomes worthless.

Nothing is more terrible than activity without insight.

Thomas Carlyle

Beginning project work without a plan is assuredly going to lead to problems. Try putting together the multitudes of pieces of assemble-it-yourself shelving without reading the directions. Using the picture on the box, you might come close. You'll probably end up with a few pieces left over, leaving you wondering if any are critical to the structural integrity of the shelves. You also get a few pieces together in the wrong sequence and have to undo them to insert another piece. All projects are much the same. Without a plan, you'll end up retracing your steps to redo work, you'll have no assurance that the work is heading in the right direction, and you don't know what product will result.

Plans succeed when you truly believe that they are valuable. There are many reasons to believe in their value:

- You and your customer will have a more accurate picture of what is going to happen as work progresses.
- You will uncover problems in the project scope and budget.
- You will better anticipate what resources you need and when you need them.
- You and your team will do a better job of thinking ahead as a result of working together to prepare the plan.
- You will be better able to anticipate problems and react to them.
- You will create a better understanding of the project by your client and your team.
- You will improve coordination and communication.
- You will build team and individual commitment by mutually identifying responsibilities and deadlines.
- You will have a basis to monitor and control project progress.

Your chances of producing a quality project on time and within budget are increased dramatically if you *properly* prepare a plan.

What are the keys to *properly* preparing a plan? They include clearly defining project goals and objectives, involving the people who will do the project work in preparing the plan, clearly defining checkpoints along the way, dividing the project into tasks that are understandable and that can be managed, taking enough time to do it right, including a mechanism and a clear intent to keep the plan updated as the project progresses, considering the inevitable problems, and including appropriate contingencies. The resulting plan must be clearly stated and effectively communicated. All of the information generated in the planning process must be presented in an easy-to-read and -use format.

If you get too close to people, you catch their dreams.

From the movie "Tucker"

The principal benefit from planning comes from engaging in it. *The process of participation in the planning effort is perhaps the most important product.* As a result of participation, the people on the project team become more committed to implementing the plan.

Let's turn now to the key steps involved in a successful plan.

DEFINE THE PROJECT OBJECTIVES

Our plans miscarry because they have no aim. When a man does not know what harbor he is making for, no wind is the right wind.

Seneca (4 B.C. to 65 A.D.)

It is essential to set clear objectives for the project. You plan a project by starting with the end result that you want and plan backward to the beginning. Even the most expert marksman can't hit a target that they can't see. How can you expect your project team, no matter how expert they are, to achieve an objective that they don't understand? As obvious as this concept seems, many projects run into difficulty or fail simply because the objective is not clearly defined and articulated. There are some pitfalls, even when there is an intent to define the project objectives.

Customer Has Unclear Picture of Needs

Management by objective works if you know the objectives. Ninety percent of the time you don't.

Peter Drucker

The end-user of the project (usually your customer) often has only a sense of their needs without fully understanding them or their implications. They may have a clear idea of what they don't want, but not of what they want. Many times, it seems that they are not sure of what they want, but think they will know it when they see it. The user's fuzzy perception of what they want from the project is a major pitfall to avoid. If you base your project plan solely on what your customer *says* they need, you are not likely to produce a result that meets their true needs.

People are usually more convinced by reasons they discovered themselves than those found by others.

Blaise Pascal

You and your project team must work closely with your customer to help them identify their needs. There are several benefits that result from this process. Your team will develop a thorough understanding of the user needs and will then be better able to plan the project. The customer becomes educated about the implications of meeting their needs and will become more committed to accepting and using the project output. Before beginning your project plan, you and your team should have a dialogue with your customer about their needs. Ask open-ended questions, such as: "What needs is this project addressing for you?" "What do you see as the biggest issues?" "What are your time constraints?" "What would the ideal end-product look like to you?"

After your discussions with your customer, write down the project objectives as you see them. The customer can then agree and tell you to proceed or can tell you where they disagree. You may well go back and forth for a few cycles. This iterative process will move you closer and closer to a mutual understanding and acceptance of the project objective. This isn't the end of the process of understanding your customer's needs. It's the beginning because their version of their needs may change as the project progresses.

Premature Solutions

Difficulties . . . are mainly due to a very simple cause: namely to attempt to answer questions without first discovering precisely what *question it is which you desire to answer.*

George Edward Moore

If you try to short circuit this process, you may come up with answers before you've even asked the right questions. It takes patience because your customer may have prematurely formed ideas about solutions before fully examining or understanding their own needs. Of course, you may have the same tendency when the customer strongly states their perception of the solution. For example, a customer may call an architect and say that they want to add a bedroom, a recreation room, and a bathroom to their existing house and direct the architect to proceed. The architect could, of course, proceed immediately to gather data on the existing house and draw the plans. If the architect takes the time to quiz the customer on the basic needs to be addressed by the project, a different solution may result. Maybe the right answer is a totally new house or modifying the existing structure without adding any new structures. Continue to ask questions until you are sure that the real needs of your customer are identified. Only then can the project objective be accurately defined.

Establish Priority of Needs

It is also important in the dialogue with the customer to determine the priority of their needs. The customer may have the noble but simplistic objective of a project that will result in a safe facility of high quality at low cost, to be built by a fixed completion date. The objectives may simply be incompatible and unrealistic. The situation can be made worse if the customer imposes other demands, such as use local subcontractors, as the project proceeds. It is important in your objective-setting dialogue with your customer to establish the relative value of each of their objectives. This knowledge will be extremely useful in planning the project and in making decisions as the project proceeds. Based on your discussions with your customer, prepare a list of their priorities as you understand them. For example, the customer's apparent priorities might be, in descending order:

1. Maximize construction and operations safety.
2. Get a quality, reliable facility.
3. Minimize operating costs.
4. Keep capital costs within budget.

5.   Complete project on schedule.
6.   Use local subcontractors.

Have your customer re-rank any misjudged priorities until there is a clear picture of the relative priorities. Establishing the priority of needs forces you, your team, and your customer to make choices about what is most important. The process will also make a valuable contribution to clearly defining your customer's true needs.

Distorting the Customer's Needs

The chief cause of problems is solutions.

Eric Sevareid

It is human nature to tend to alter your customer's needs to more closely match your perceptions of what their needs or the solution to their problem should be. This often subconscious process can occur in several ways. First, there can be a tendency to give the customer more than they truly need because of your superior knowledge of the subject. For example, the customer may have a need for a simple, manual control system for some rarely used emergency pumps. Because you are aware of the state-of-the-art control systems, you may feel that the customer needs a computerized system complete with telemetry of the system status to the customer's main office building. The iterative process of needs analysis described earlier is effective in catching such major differences in perception of needs before work begins.

To a four-year-old boy with a hammer, all the world is a nail.

Anonymous

There is also a danger that you will apply your favorite solution to the customer's needs, whether appropriate or not. In the process, the project objective becomes distorted. We have seen some consultants that apply the same solution regardless of the customer's needs. As a result, the customer ends up with a project that may be overly complex and unduly costly, and ends up being underutilized. Again, the iterative needs analysis process protects against this result, especially if you involve a group of people with different backgrounds.

Ambiguous or Imprecise Language

Don't assume that you'll always be understood. I wrote in a column that one should put a cup of liquid in the cavity of a turkey when roasting it. Someone wrote me that "the turkey tasted great, but the plastic cup melted." So now I say: "Pour a cup . . . "

Heloise

Even after the iterative needs analysis process, there is a danger that the written statements of the project needs will be unclear. You are in trouble if two people look at the same statement of the objective and disagree on its meaning. For example, there may be agreement that the project involves painting the customer's house brown. Of course, there are many shades of brown. Ask the two people to pick a "brown" paint and you'll get two different paints. The ambiguity could be removed by specifying a certain paint identification number.

The difference between the right word and the almost right word is the difference between lightning and the lightning bug.

Mark Twain

The customer's objective may be a marketing study for a new product. Such a project deals with a lot of abstractions, as opposed to tangible objects that can be drawn or precisely specified. In this case, help your customer see what the end-study can look like to clarify their objective. Ask them for a similar study that they've seen and liked. Show them a marketing study that you have done for another customer as an example. They may not have the experience to know exactly what the final study should look like, having instead a general knowledge that they need to know the potential market for their product. By helping them to visualize the final study by showing them examples, you can clarify the project objectives. They will likely identify parts of other studies that they like and dislike, removing much of the ambiguity about what they need.

There are also some key phrases to look for in a definition of the objective that are likely trouble-causers due to their inherent ambiguity. Terms such as "optimum," "maximum (or minimum) extent possible," "approximately," "at least," "include but not necessarily limited to," and "nearly." If there is a term that is open to varying interpretations, it doesn't belong in the definition of the project objective. Dig deeper with your customer to find out their true needs. If they say they want you to evaluate "at least" six alternatives, dig deeper to find out exactly how many they feel would be a reasonable basis for the project and why.

Use of Jargon

Don't, Sir, accustom yourself to use big words for little matters.

Samuel Johnson

Your customer is not likely to be an expert in your field. The objective cannot be clearly stated if it uses language that the customer cannot understand. Avoid the use of jargon in stating the objective. Put the objective in terms or an analogy that the customer uses and understands.

What The Objective Should Look Like

Don't waste your time trying to control the uncontrollable, or trying to solve the unsolvable, or thinking about what could have been. Instead, think about what can be if you wisely control what you can *control and solve the problems you can* solve *with the wisdom you have gained from both your victories and defeats in the past.*

David Mahoney

We've devoted several pages to issues related to establishing the project objective because it is vital to project success. It establishes your direction. It focuses the energy of you, your team, and your client, and keeps all of you pointed in the same direction. It creates commitment, and gives you a way of knowing when you are done. To be effective, the objective should have several key characteristics:

- **Be specific**—Anyone with basic knowledge of the project subject should be able to read the statement of the project objective and understand it.
- **Be measurable**—The objective needs to be measurable in terms of schedule, scope, and budget. Some project objectives are more difficult to measure than others, but all must be measurable if the project team is to have a sense of direction. Projects with intangibles take the most effort to establish measurable objectives for. One bank wanted to increase the perception of its customers that the bank personnel were friendly. To measure the result of the project, the bank management decided to count the number of comments between the customers and the bank personnel unrelated to the transaction (e.g., weather, sports, compliments on customer's appearance). The staff all bought into the project. After five weeks, the number of such comments had increased and a follow-up survey of the customers showed that they felt the bank was more friendly. If you work at it, you will find some way to measure the objective. Without such a measure, you simply don't have a clear objective.
- **Be realistic**—Objectives that cannot be accomplished will cause your team to lose its motivation. Certainly, some stretch is desirable to challenge the team, but it should not be a stretch to the breaking point. It is crucial that the availability of personnel, other resources, and timing be discussed and considered in setting the objective. Don't start a project with an impossible deadline or inadequate staff. Also, be realistic about how your skills and capabilities match the project needs. Stick to things that you have some experience in. Rarely is it realistic to expect to learn a new area during the course of a project.
- **Be in agreement**—You and the customer must be in agreement on the project objective. By involving your team in defining the objective, they will also be in agreement, building their commitment to the project.

- **Be assignable**—The objective must be capable of being divided into tasks for which, ultimately, an individual will be responsible for. This allows for accountability and project control.

The world stands aside to let anyone pass who knows where he is going.

David Starr Jordan

Once the objective is clearly established, you need to keep everyone focused on it by distributing a written statement of the objective to the team and by constantly reminding them of the objective. Put it in a summary form that can be posted for all to see each day. If you walked through any of the City of Austin, Texas facilities in 1991, you would see the statement of objectives for the city work force posted in prominent locations (see Table 7-1). Constantly remind the team members that this is the objective that their efforts are to be focused on. If they are doing something that does not advance them toward the objective, it's time to reassess and align their activity with the objective.

Fanaticism consists of redoubling your effort when you have forgotten your aim.

George Santayana

TABLE 7-1   Example Public Statement of Objectives

CITY OF AUSTIN—MANAGEMENT PLAN
VISION
We are committed to providing responsive, quality services that contribute to the well-being of our community. We will transmit this City greater, better and more beautiful than it was transmitted to us.

VALUES
We will honor the public trust through . . .
- Respect and dignity for our customers.
- Open and honest communications.
- Respect and care for the environment.
- Positive action and innovation.
- A workforce selected with care, treated with respect, and rewarded for performance.
- Teamwork.
- Responsible use of public resources.

GOALS FOR 1990-91
- We will focus on customer service.
- We will invest in the workforce.
- We will live within our means.

On this date, November 6, 1990, we, the undersigned, personally commit to this vision, these values and to achieving these goals: (signed by City of Austin employees).

PLANNING YOUR TEAM

There are only three rules of sound administration, pick good men, tell them not to cut corners, and back them to the limit; and picking good men is the most important.

Adlai E. Stevenson

The people working on your project will determine its success or failure. As the objective definition process proceeds, you may find that the nature of the team needs changing as well. Naturally, you'll be looking for the best qualified people who have the time available when you need them. Usually, projects are carried out within an organization concurrently with several other projects. You inevitably end up balancing skills and time availability to assemble the best team possible. It is better to be realistic about people's availability when structuring your team. It's of no value to have the world's greatest computer programmer on your team if she or he is not going to have time available when you need it.

If you assemble more then seven people on a research project, you can be assured that research effectiveness will decrease.

Tom Peters and Bob Waterman

Use as few people as possible on your team to produce the bulk of the project work. The potential for communications problems increases rapidly as the number of people on the team increases. For example, a team with 20 members has 190 potential communication channels. With 5 members, there are only 10 potential channels. Offsetting the risk of a large number of participants is the fact that a larger, more heterogeneous group offers a greater diversity of views. Bring in senior experts for periodic reviews, to reduce the limitations of views represented on your project team. Generally, your team will be more efficient with a few people working full-time on your project, as opposed to many people working part-time on the project. If your organization has several projects underway, each with a lot of people working a little bit on each one, take a serious look at restructuring their assignments. Wherever possible, reduce the number of projects that each is working on so that they can spend most of their time on a single project. If you're working within an organization, it's prudent to involve the upper management of the organization in this step as well as your team members.

DIVIDING THE OBJECTIVE INTO TASKS

A vision without a task is but a dream./A task without a vision is drudgery. A vision and a task are the hope of the world.

Unknown, from Church in Sussex, England (1730)

In order to do the project, you need to divide the objective into tasks, each with its own objective, hours, schedule, budget, defined inputs and outputs, and responsible individual. The task descriptions tell who will do what, when they will do it, how much they will spend to do it, and how progress will be measured.

Your goal should be to use the *minimum* number of tasks necessary to describe the work. Some people seem to think that having a large number of tasks leads to clearer definition and more control. Instead, it leads to a loss of control because the people working on the project will not be able to track how their time is divided among a multitude of tasks. You won't have time to track numerous minor tasks either. The task outline should be as simple as possible so that it can be easily tracked and updated.

In defining the tasks, use the *same* list of tasks from the scope of services in the project contract, in the project schedule, and in the project budget. By being consistent, it will be easier to track progress and relate it to your original plan and contract. It will also be easier to document any changes in scope as the project proceeds.

The essential elements of each task are:

- A defined scope.
- Clear definition of the end-product of the task (e.g., a task report, a prototype, drawings, specifications).
- A start and finish date.
- If the task is long enough or complex enough, milestones to check progress.
- Required inputs.
- A budget.
- Who is responsible (the task manager).
- Personnel expected to work on the task and the amount of effort expected from each. Even if not used for cost control, it is prudent to allocate the hours for the subtasks, to get agreement and to increase team member awareness of their efforts.

Table 7-2 is a sample task description. Each task should be defined in an iterative process with your team members. Involving the team members in developing the task descriptions is essential. *Without involvement there will be no commitment.*

The two qualities which chiefly inspire regard and affection (are) that a thing is your own and that it is your only one.

Aristotle

Everyone on the team needs to understand and accept how their responsibilities contribute to achieving the project objective. Get the people who are going to do the work to plan the work. After all, it is going to be their task to do, not yours.

TABLE 7-2   Example Task Description
Task 204—Long-Range Plan

Budget	Task Leader	Culp	40 hours
		Smith	40 hours
		Williams	24 hours
		Jones	8 hours
		Sr. Engineers	60 hours
		Project Engineers	120 hours
		Secretarial/Clerical	40 hours
	Amounts	Labor	$34,500
		Expenses	2,500
		Total	$37,000
Schedule	Start Date	May 1, 1991	
	Finish Date	September 1, 1991	

Task Summary   Based on regional plant study, determine sizing and timing of future plant expansions, assuming regional plant isn't built. Develop a conceptual plan for these expansions to year 2015, showing inclusion of short-term improvements planned in Task 203. No cost estimates for long-term improvements are to be included. Prepare a task memo.

Details and   1. Contact client to get population and flow   (April 15)
Schedule   projections from regional study

2. Define future flows to 3 treatment plants   (May 1–May 10)

3. Using layouts from Task 203, define long-term improve-   (May 10)
   ments at each plant, size each process, and add to layouts

4. Prepare drawings of layouts   (July 15–July 30)

5. Prepare narrative text that describes basis for   (July 1–August 1)
   long-term improvements, alternatives considered,
   how short-term improvements were interpreted

6. Typing of draft memo   (August 1–5)

7. Submit draft memo to client   (August 6)

8. Receive client comments   (August 20)

9. Prepare final task memo   (August 20–September 1)

Output   A task memo showing the long-term improvements to serve population projected in regional plant study.

Input   Definition of short-term improvements in Task 203   (May 1)
Population and flow projections from regional plant study.   (April 15)

They should know more about it than anyone else. Generally, people are motivated to perform project work to the degree that their activities serve their personal values and goals. By allowing them to shape their task descriptions, they have an opportunity to relate their work to their values. Their sense of ownership goes up, as does their motivation. When you own something, you take better care of it. When was the last time you waxed a rental car?

After giving the team members some time to consider the project objective, bring them together for a workshop to define the project tasks. This approach has several advantages:

- Everyone learns who they will be working with and has a chance to get to know them.
- An early focus on planning is provided.
- Input is obtained from the team, with the potential for synergism as people with different backgrounds and perspectives discuss the tasks.
- Potential for later conflicts is reduced.
- Keeps people from focusing too narrowly on their own tasks.
- You get a better feeling for each of the team members and how they will best work together.
- Inexperienced staff get help on defining their tasks, saving time and reducing estimating errors.
- The group begins to develop a team identity.

To be productive, such a workshop needs to be controlled by using the methods we discuss in Chapter 6 for effective meetings. It is a group discussion, not a technical dissertation by you on the project. Each participant should have a thorough understanding of the project objective and his or her potential area of activity before the workshop. In addition to working on task definitions, the group can discuss team member roles, communication procedures, how decisions will be made, how project changes will be made, and the coordination of resources. This does not relieve you of the responsibility of thinking it all out beforehand. If you don't, the meeting will likely drift and will not inspire confidence in you. By being prepared, it will be easier to focus the meeting on the key issues.

There are two ways of meeting difficulties: You alter the difficulties or you alter yourself to meet them.

Phyllis Bottome

It is also important for the team to brainstorm major potential problems and what actions might best address these problems. For example, the team might identify that obtaining a federal permit could cause a major delay. The action steps could include an early definition of permit submittal requirements, early collection of the needed information, and assignment of an individual to monitor the progress of the application.

Action might also involve adding someone to your team that has specialized skills to resolve the problem. Involvement of others may be increased or decreased as problems are considered. By considering alternative pathways to the project objective, the team will be better equipped to cope with problems when they arise.

After the workshop, reduce all the input into a written description of the tasks, including the definition of the essential task elements listed earlier. Circulate the written material to the team for comment. Use the back-and-forth process until there is agreement. This may require an intense effort on your part to accomplish this in a reasonable time, with a lot of one-on-one discussions after the workshop. The rewards in terms of fewer problems as the project proceeds justify the effort. When tasks are understood and are within the capabilities of the team members, there will be a high degree of confidence and commitment. By throwing the project plan open to discussion, you foster the involvement that leads to team commitment. It is still your responsibility as project manager to pull all the pieces together into a coherent, rational plan.

PROJECT SCHEDULING

Lose and neglect the creeping hours of time . . . and know what 'tis to pity, and be pitied.

William Shakespeare

When the objectives and tasks are defined, you need to develop a project schedule. As you work with others on your team to define the schedule, remember that people's tolerances for risk varies substantially from person to person. Some will naturally give you optimistic schedules, while others will be very cautious, adding in their own contingency time. Disagreements over planned task durations often relate to differences in tolerance for risks, rather than basic disagreements over the time needed to do the work.

As you assess the reasonableness of the time for each task, you must evaluate it in terms of the capabilities of the people who will actually do the work. A common pitfall is to evaluate the time required as if *you* were going to personally do the work on the task. The people doing the work may be either more or less expert than you. Adjust your estimate of the time required in light of their abilities and your past experiences with similar tasks. If the task is something that you've been involved with many times in the past, you have a good basis for evaluation. For example, someone projects a time of 9 hours to play 18 holes of golf. You know from experience that the time is usually 31/2 to 5 hours, depending on the golfer's skill, physical condition, and how crowded the course is. You'd have good reason to question the estimate.

It is a bad plan that admits of no modification.

Publilius Syrus (first century B.C.)

No matter what scheduling technique you use, include contingency time. All projects deviate from plan at some point. If there is no contingency time, the first missed task deadline creates an impossible chain of other task deadlines. Team de-motivation results.

There are a large number of scheduling methods in use, each with its own advantages and disadvantages. The scheduling method best suited for your project will depend on the scope and complexity of the project, the number of tasks, the interrelationship between tasks, the duration of the project, and the number of people involved. The methods range from simple to very complex. We'll briefly review the advantages and disadvantages of each and how they affect the people working on the project. There are a multitude of books on the mechanics of these scheduling methods.

Milestone Charts

To act intelligently and effectively, we still must have a plan. To the proverb which says, "a journey of a thousand miles begins with a single step," I would add the words "and a road map."

Cecile M. Springer

Milestone charts are probably the simplest scheduling method. In its basic form, this method involves identifying and tabulating the completion date and person responsible for each task. The milestones are like markers along a road to let you know that you are headed in the right direction and at the right speed. They provide an indication of whether you need to change your speed and how much time is available to finish the project. They are tied to major events (e.g., submittal of a report outline, submittal of the draft report) that are dictated by the overall project scheduling needs. Remember our imaginary trip from New York to Los Angeles in Chapter 5? If you were driving, you might set up milestones such as to be in Pittsburgh by the end of Day 1, Springfield by the end of Day 2, and so on. Reaching each city on the right day would indicate that you were on course and getting closer to your objective of Los Angeles. If you found yourself in Miami, you'd know that you were seriously off course.

Milestones help motivate and focus people on your team. They are like a map that helps them pinpoint their location on their journey and keeps them from straying from the road on the way to the project objective. They provide a means to monitor progress and provide feedback. Each person can keep track of their own progress because they have a copy of the map. A milestone chart for a project might look like:

Task Description	Responsibility	Target Date	Actual Completion Date
Task A	TJM	9/1/93	8/29/93
Task B	RMB	10/4/93	10/7/93
Task C	HHB	10/20/93	
Task D	TJM	11/15/93	

The milestones should be a few major events critical to the project schedule. Don't include so many that they becomes inchstones to stumble over.

Although simple and useful as a general map, milestone charts have the disadvantage of not showing when the task should begin or its relationship to other tasks. They have advantages in that they are low in cost to prepare, easy to update, and easy to understand. They are often used as a summary when more advanced scheduling methods are used for project control.

Bar (or Gantt) Charts

Henry Gantt developed this commonly used method in the early 1900s while working for DuPont. Figure 7-1 illustrates the concept. The sample project was divided into five tasks. When the chart was prepared, five open bars were drawn to represent the time span for each task. As the project proceeded, a portion of each open bar representing the percentage completion of each task was shaded in. In the example, Task A is completed. Task B is about two weeks behind schedule. Task C is about two weeks ahead of schedule. No work has been started on Tasks D or E. Bar charts are easy to construct and understand. Similar to a milestone, they can give the project team a quick overview of project status. In a glance, you can tell where work is ahead or behind schedule.

Bar charts do have disadvantages as well. It is often difficult to accurately assess the percent completion of each task. Consider a project of digging ten post holes. You have the first nine done. Looks like you're 90 percent complete. Then you hit a boulder on the tenth hole. You either have to realign the fence or bring in a backhoe. Even if the current status is accurately known, there is no information on the time required to finish the task. The chart ends up being an approximation of the task status. The dependence of one task on another is not shown. Does Task B have to be complete before Task D can start? We can't tell from the bar chart. Bar charts are very useful as a snapshot of project status, but not so useful as a planning tool to make sure that things happen properly in the future. They can be useful in summarizing project status for your customer. They can also give you an indication

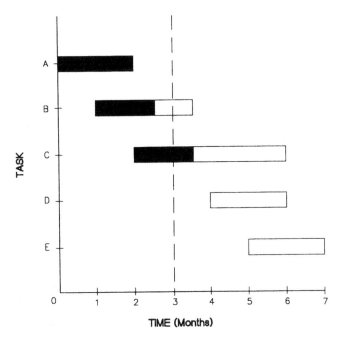

FIGURE 7-1   Example bar (Gantt) chart.

of where there may be resource imbalances. For example, based on Figure 7-1, maybe you could divert some resources from Task C, which is ahead of schedule, to Task B, which is behind schedule.

Flow Charts

In the early 1950s, two widely used methods that chart the flow of the project schedule were developed: the Critical Path Method (CPM) and the Program Evaluation and Review Technique (PERT). They are often referred to as network diagrams as well. There are volumes of papers and texts on the conventions and nomenclatures that are involved in preparing these charts to which we refer the reader.

The development of PERT began in 1958 when the Navy had the goal of producing the Polaris Missile system in record time. Durations of similar projects were averaging 40 to 50 percent greater than estimated. Until then, scheduling systems were traditionally based on the concept of a fixed time for each task. In the PERT system, estimates were made for the most likely time, the most optimistic time, and the most pessimistic time. This range provided a measure of the uncertainty associated with the actual time to complete an activity.

The expected time of a task is defined as:

[optimistic time + 4 (most likely time) + pessimistic time]÷6

This calculation may be useful to you for critical activities with which you have little experience. It is possible, with PERT, to develop probabilities that a project will be completed on time. PERT, as an effective project manager should, recognizes that the concept of a definite completion time for a project is misleading. Statements of a possible range of times and the probabilities associated with each are more realistic.

CPM resulted from a joint effort by DuPont and Remington Rand Univac in 1956–1959 to determine how to reduce the time required to perform routine plant maintenance and construction work. One goal was to determine the optimum trade-off between project duration and project cost. CPM treated the task performance times as a single value, as opposed to the PERT approach. The main feature of CPM was its ability to arrive at a project schedule that minimized total project costs.

CPM and PERT were developed concurrently by two different groups, each unaware of the efforts of the other. Both are now used for a wider range of applications than for which they were originally developed.

Although substantially more complex than bar charts, flow charts provide a great deal of useful information on identifying and managing the sequence of task work. The sequencing of tasks is important. Imagine trying to build a long pipeline by starting at several places scattered along the alignment. Fitting together all the pieces would be more difficult than if you started at one end. Sequencing of tasks also affects your ability to monitor and control progress. In arranging the sequencing of tasks, consider carefully what inputs are needed from other tasks and what tasks are dependent upon outputs from the task under consideration. Flow charts show which tasks depend on the output of other tasks. With computer software now available, it is relatively easy to sequence hundreds of activities. They are also used to add up the time estimates along each path of activities to the project objective. The longest path is the "critical" path since all activities on this path must be completed before the project objective is reached. Although flow charts are excellent tools for project control, they tend to be too complex to readily show the status of the work. Thus, a bar chart is often used to summarize project status and a flow chart for project control.

Flow charts are useful for "what if" analyses. You can program a potential delay in one task, and use the computer software to predict the net result on the overall project schedule and the schedule of other individual tasks.

Flow charts offer several advantages:

- Make it easier to see how time and resource constraints affect the project.
- Reduce the risk of overlooking necessary tasks.

- Improve coordination between tasks by forcing consideration of task interrelationships.
- Encourage a team feeling by demonstrating the interface and interdependence of various tasks.
- Focus management attention on the few activities that have the most impact on the project schedule.

Avoiding Scheduling Pitfalls

The future is arriving on an express train, and it might arrive before we're ready.

Jack Anderson

Get the team committed to the schedule by having them involved in preparing the schedule. Keep the schedule up to date. No matter which method is used, it is worthless if not updated. Every project undergoes change as it proceeds. These changes must be incorporated into the schedule if it is to have any value.

Often, project reviews are included in the schedule, but no time allowance to evaluate and implement the findings of the review is included. Similarly, allowing no contingency time is a major pitfall—all schedules slip. The project schedule is of no value unless it is used. To be used, it should be easily understood, practical, pertinent, and up to date. Avoid project schedules that are so cumbersome and complex that they can be used only by a scheduling specialist.

Be realistic about the time required in the final stages of the project. People often underestimate the time required to incorporate comments and finish up all of the details needed to complete a project. Also recognize that you don't have control over the time it may take your customer (and local, state, or federal regulatory agencies if they are involved) to review your project before it can be completed. It is prudent to schedule your project completion tasks from the date you receive the final review comments.

PROJECT BUDGETING

Money is like a sixth sense without which you cannot make a complete use of the other five.

William Somerset Maugham

Much like Maugham's sixth sense, you can't make use of your project objective and schedule without an adequate budget. Don't attempt to prepare the budget until you have completed the project scope and schedule. There is no way to accurately estimate the budget without both the scope and schedule because they determine the resources needed. Seems obvious, doesn't it? Well, we've seen more than one case where preoccupation with the financial aspects of a project caused the budgeting step to be taken out of this logical sequence. It doesn't work.

Budgets are not the same as cost estimates. The Trans-Alaska Pipeline was first estimated to cost $900 million dollars in 1968. When completed in 1977, the costs totalled $7.8 billion. Conditions can change (i.e., inflation was rampant), and knowledge can increase (little was known about the effects of the arctic and environmental concerns on pipeline costs in 1968) from the time of estimates to the time of project initiation. Don't allow earlier estimates to determine a project budget. Only after the project is defined and scheduled is it time to do an independent budget.

Budget: Telling your money where to go instead of wondering where it went.

C.E. Hoover

There are two basic philosophies about preparing a budget: the "top-down" and the "bottom-up" approaches. In the top-down approach, the overall project cost is projected based on some experiential factor, such as dollars per square foot of building or a percentage of construction costs for design services. Such an estimate is not directly related to the level of effort needed for the project. The task budgets are then developed by working backwards from the top-down total. Although this may be a useful reality check on a budget, it is not a good way to prepare a project budget. It does not provide good planning and control information because the resulting task budgets may not have any relation to the required effort.

In the bottom-up approach, the hours of effort for each task are estimated and multiplied by the labor rates of those who will do the work, to determine the labor costs. Material costs and other expenses are also estimated for each task, as are the appropriate contingencies for each task. This approach has the advantages of forcing the project team to consider the requirements of each task, providing a meaningful basis for project control, and being a good basis for negotiations of project costs with your customer. The detailed tasks, scopes, and budgets will give your customer a clear picture of the cost and time implications of various aspects of the project. If the customer wants to alter the scope, they have a basis upon which to judge the effects. If the customer has champagne tastes and a beer budget, the detailed task scopes and budgets often result in appropriate increases in the project budget.

Compare your bottom-up estimate with a top-down estimate to determine reasonableness. If the top-down estimate is considerably higher, look for costs you may have missed not directly related to the tasks (don't forget project administration costs!) that you've identified, or for some unjustified simplifying assumptions that have been used. If the bottom-up estimate is higher, be sure that you haven't used multiple contingencies on the same work items.

As in the case of developing the schedule, involve the team members responsible for doing the work in estimating the cost. As was the case in estimating time, there is no way to be certain of the cost of each task. It is prudent to include a contingency allowance on a task-by-task basis because the risks vary from task to task. A task of high certainty should include a very low contingency—maybe none is needed in the task of typing the report. There may be tasks involving great uncertainty over the amount of labor required (such as obtaining a permit from the Corps of Engineers or the Environmental Protection Agency), where a 50-percent contingency is appropriate. Be careful not to add contingency upon contingency because multiple contingencies can become unreasonable. Apply appropriate contingencies to each task based upon its uncertainty, rather than a blanket contingency applied to all tasks. Use the total of the task contingencies as the single project contingency. Transfer funds from the project contingency back to individual tasks as the need arises.

Don't plan costs in any greater detail than you will be receiving for the project cost information. It's pointless to budget costs for subtasks for which no separate cost information will be reported, or to budget weekly costs if only monthly cost reports will be available.

There are several points of caution to consider in your budgeting process. Make a separate allowance for inflation. No one can say for sure what inflation will actually be. By identifying a separate item, you will have a basis to request a change from the client if the actual inflation rate differs. The people who end up doing the work may not be the same ones that you anticipated. The person who could have done the work in 80 hours may be gone, only to be replaced by someone who will take 120 hours. Don't be timid about contingencies. If the contingency is inadequate, you'll have to negotiate a contract change every time a problem arises. Project administrative costs are often underestimated. Monthly reports, meetings with the client, and coordination of subcontractors and team members often seem to take more time than estimated. Project management costs often are 10 to 15 percent of overall project costs, sometimes even more. Costs to close out a project are also often underestimated. Some of the special concerns about ending a project are discussed in the next chapter.

As with schedules, involvement of the project team in preparing the budget is essential. Remember—no involvement means no commitment.

PUTTING IT ALL TOGETHER

When schemes are laid in advance, it is surprising how often the circumstances fit in with them.

Sir William Oster

You have the pieces now—a clearly stated project objective, a schedule, and a budget. It's time to integrate the pieces into the project plan. As you do, remember that the plan is a dynamic tool that will allow you and your team to cope with changes in an orderly manner. Assuredly, some things will go wrong or conditions will change as the project proceeds. There is no such thing as a perfect plan that anticipates every problem or condition. Recognize that problems and changes will arise and have a plan to minimize impacts.

It will be useful to integrate all the critical project information into a project manual. Although organization of the manual will vary depending on the project, the following offers some typical sections:

- Project description.
- Statement of project objective.
- Project team—include addresses, phone numbers, and FAX numbers for each team member if they are at different locations.
- Detailed scope—include the scope from the project contract. Don't paraphrase the contract scope because you may mislead someone.
- Task responsibilities—list person responsible for each task.
- Schedule.
- Budget.
- Sample cost reporting sheets with related explanation of content.
- Invoicing procedures.
- Reporting procedures.
- Procedures for changes in scope, budget, or schedule.
- Quality control and management procedures.

Issue revisions to the plan as changes occur. Be sure that everyone on the team gets every revision promptly.

Remember that the plan is nothing more than a tool. It is a very important tool, but is a means to an end—the project objective—and not an end unto itself. Combine your project plan with the person-centered skills and procedures described in the other chapters and you are on your way to project success. Involving your customer and your team throughout the planning process creates a plan that belongs to all involved. The ownership of the project plan by the people involved with the project is critical because . . .

People make a plan work, plans alone seldom make people work.

Confucius

References
1. Randolph, W.A., and B.Z. Posner. 1988. *Effective Project Planning and Management. Getting the Job Done.* Englewood Cliffs, New Jersey: Prentice Hall.
2. Rosenau, M.D. 1984. *Project Management for Engineers.* New York: Van Nostrand Reinhold.
3. Graham, R.J. 1989. *Project Management As If People Mattered.* Bala Cynwyd, Pennsylvania: Primavera Press.

4. Frame, J.D. 1987. *Managing Projects in Organizations.* San Francisco: Josey Bass, Inc.
5. Kerzner, H. 1989. *Project Management. A Systems Approach to Planning, Scheduling and Controlling.* New York: Van Nostrand Reinhold.
6. Burstein, D., and F. Stasiowski. 1982. *Project Management for the Design Professional.* New York: Whitney Library of Design.
7. Olde, W.T., and M. Peralta. 1989. The proactive execution plan. *Project Management, A Reference For Professionals,* edited by Kimmons, R.L., and J.H. Loweree, p. 335. New York: Marcel Decker, Inc.
8. Pincus, C. 1989. A workshop approach to project execution. *Project Management, A Reference For Professionals,* edited by Kimmons, R.L., and J.H. Loweree, p. 349. New York: Marcel Decker, Inc.
9. Moder, J.J. C.R. Phillips, and E.W. Daves. 1983. *Project Management with CPM, PERT and Precedence Diagramming.* New York: Van Nostrand Reinhold.
10. Hartley, K.O. 1989. The project schedule. *Project Management, A Reference For Professionals,* edited by Kimmons, R.L., and J.H. Loweree, p. 365. New York: Marcel Decker, Inc.

8

Monitoring and Controlling
Project Progress

One machine can do the work of 50 ordinary men. No machine can do the work of one extraordinary man.

Elbert Hubbard

People do projects. Systems and procedures do not. Control of the project depends less on the technical beauty of your control system than it does on the way you use the information and the way you relate to your team. Monitoring and controlling projects requires you to communicate with people, measure their progress, and get their commitment to course corrections. A multitude of systems are available to track costs, staff hours, task interrelationships, and so forth. Systems organize information about the project, but people plan, control, and implement the project.

Some managers get confused. They try to manage numbers and report people.
To be successful, you have to manage people and report numbers.

William Hoglund

You cannot manage a project with data and reports alone. Reports are a one-way communication and do not provide the exchanges between people necessary to accurately evaluate the status of a project. You must integrate communication skills, experience, cost and schedule information, and technical expertise to assess the status of a project and make appropriate adjustments. Rarely will technical expertise alone compensate for the lack of people and communication skills.

Estimating the progress of each project task is the most important step in monitoring and controlling project progress, and it is largely a nonmathematical step that relies on your experience and judgment. A successful project manager seeks and obtains accurate input, support, and advice of the project team. One way to get this input is management by walking around (MBWA), a crucial element in monitoring a project. Personal contact with the team members builds rapport, while also giving you the opportunity to personally see and discuss the status of the work. Major project cost overruns and schedule failures can occur because the project manager does not personally dig into the status of the work. There is a tendency for those working on a project to be optimistic about the time and effort to complete their work. It is critical that you personally discuss the work underway and form your independent opinion. By focusing on the people doing the work, you gain the knowledge needed to monitor and control the project.

PURPOSE OF PROJECT MONITORING AND CONTROLLING

You can boil down the purpose to two words—"no surprises." You measure progress toward the project objectives, evaluate what needs to be done to reach the objectives, and take appropriate actions. The process is a continuous one to detect and resolve potential problems so that the project stays on schedule, within budget, and produces the end result desired by your customer.

PROJECT ELEMENTS THAT CAN BE CONTROLLED

Thinking well is wise; planning well, wiser; doing well, wisest and best of all.

Persian Proverb

There are six basic elements of a project that you can control:

1. **Scope**—Every project begins with defining its scope. Even if it is precisely defined at the start of the project, you cannot assume that everyone working on the project will be working toward the same scope throughout the project. The dreaded problem of "scope creep" can be avoided only by effective and frequent communication of the scope to the project team. Your continuous assessment of how their efforts relate to the scope is essential. Scope change is common. These changes must be clearly understood by the team so that they are all working toward a common goal.
2. **Time**—There is usually a finite amount of time in which the work must be done. Unless your customer agrees to a change, you can't control the time available once the schedule has been agreed upon. You can control the time at which the work on a given project task is done relative to the overall project schedule and relative to other tasks. You may also be able to control the amount of time required to do a task by the resources you assign.

3. **People**—The number and type of people working on the project, and when they do it, are critical elements. If it is going to take 30 staff days of work to roof a building, you can't expect to do it in four hours by putting 60 roofers to work.

4. **Costs**—Obviously, cost control is important. No matter how technically successful the project is, you won't get many chances to do more projects if your costs consistently exceed the budget. Although important, tracking costs is all too often confused with project control. Analyzing the costs in relation to overall progress and taking appropriate action constitutes control—tracking alone does not. There is little solace to be gained from tracking costs, finding that the rate of expenditure is right on plan, and discovering when the funds are exhausted (right on schedule!) that the work is only 75 percent complete. This sorry result can be avoided by keeping in touch with the people doing the project work to assess their progress in relation to cost.

5. **Quality**—We've devoted an entire chapter (Chapter 9) to managing quality. There is no point finishing on time and on budget if the result won't work.

6. **Communication**—Critical communication occurs in at least five ways: from the project manager to the team, from the team to the project manager, among team members, between the project manager (and sometimes from the team members) and the customer, and between the project manager and the organization's upper management. Each of these links is important to project success and must be managed and monitored as carefully as schedule and costs. Projects planned to the Nth degree with wall-sized PERT charts showing the relationships between a multitude of tasks can fail because the project manager does not effectively communicate with the project team. If the manager doesn't really know what the team is doing, a sophisticated control system will surely fail.

MONITORING AND CONTROL SYSTEMS

Murphy's Laws:
1. *In any field of endeavor, anything that can go wrong, will go wrong.*
2. *Left to themselves, things always go from bad to worse.*
3. *If there is a possibility of several things going wrong, the one that will go wrong is the one that will do the most damage.*
4. *Nature always sides with the hidden flaw.*
5. *If everything seems to be going well, you have obviously overlooked something.*

To avoid overlooking something, a project monitoring system should be used. The purpose of the system is to provide the information needed to measure progress against the schedule, the budget, and other appropriate parameters. Prerequisite to using any system is the development of a plan to forecast progress and resource expenditures. The system is only as good as the original plan against which it is measured, as well as the parameters used for measurement. It is more common for the plan to be the weak link rather than the monitoring system. Figure 8-1 illustrates the relationship between the system that provides the information and the project control.

You control a project to be sure that the right resources are allocated to the project at the right time. The information system has to provide, in a timely manner, the type of information needed to assess the project status and to make decisions.

Interaction between you and the project team generates some of the critical information to be used in conjunction with cost and other information. Project control comes about when this information is used to make decisions and direct the use of resources to complete the project. The information is an important tool that must be used by people to coordinate and direct the efforts of other people.

It is important to get the information system in place at the earliest possible date. The information system should relate cost and schedule performance so that problems can be identified and traced to their source. To do so, it must have the ability to track significant parameters, be responsive in a timely manner, be able to predict cost and schedule for completion, and report a level of detail appropriate to the decision-making process. There is often a tendency to develop more information than is needed for project control. Don't make the simple complex. Carefully consider what information you will really need to make critical decisions and then design the information system for the project. You also need to consider what information you can accurately get. For example, breaking a project into 50 small tasks may give a false sense of accuracy. If the same people are working on

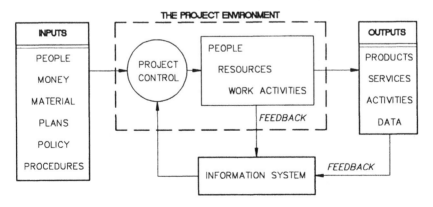

FIGURE 8-1. The project control process (based on reference 1).

several of the small tasks every day, they won't be able to accurately keep track of the hours or minutes on each task. Also, there may be no point in knowing. You could become buried in reams of inaccurate, meaningless data, while needlessly burdening the project team with reporting procedures. Consider what you'll do with the information before you generate it.

Software is now available for personal computers that could only be found on mainframe systems only a few years ago. Whether or not a computerized system is necessary depends on the size, complexity, and duration of the project. Although a sophisticated software package cannot substitute for skilled project management, it can be a help on large projects. Costs for systems can range from $300 to $3,000 for PCs to over $100,000 for mainframe systems. The most common features plan and track project tasks, resources, and costs. Task elements with their start and finish times, assigned resources, and actual data on time required and costs are entered and updated as the project progresses. The program documents the schedule and cost status against the original project plan. The software can usually generate reports in a variety of detail and formats. Many programs also allow you to try "what-if" analyses, to estimate the effects of different courses of action.

There are literally hundreds of such programs available, clearly far more than we could describe here. You'll need to evaluate the ones most appropriate to the level of detail and complexity of your project, if in fact a computerized system is needed. Talk to someone who has several months experience with the systems that you're considering. Their experiences can save you a lot of grief. Some factors to consider in evaluating project management software follow:

- What is the quality of documentation?
- Is the program user friendly?
- Does the program include graphics that can be used to quickly grasp the project status?
- Does the vendor provide support, such as program modification, information on other users, and updated programs?
- Does the program require special hardware?
- Does the program have the flexibility and capacity to cope with the amount of data involved in your project?
- Can the program be integrated with your company's information system?

You may be part of a large company that has a corporate-wide management information system (MIS) in place. Is there still a need to consider separate program information software? Yes. Large MIS systems usually do not generate the specific information useful for project management. Also, they typically generate too much information. You sift through reams of printouts to find the data you need. They are usually not responsive enough to provide the information you need when you need it. The large MIS systems are simply designed for another purpose—overall corporate management—as opposed to individual project management.

Before turning to other aspects of project control, one more reminder that gathering project information and tracking progress does not constitute project control. The information system cannot be permitted to become an end unto itself. Don't fall into the trap of thinking that all of the information you need can be viewed on your computer screen or printout. The project manager and team control the project, not the computer. You must interact with the project team to form your own assessment of project progress.

ESTABLISHING CRITERIA TO JUDGE PROGRESS

Project progress and expenditures are not necessarily related. A common error is to attempt to track project progress by plotting a graph, such as Figure 8-2, that compares actual costs to date versus planned costs. From Figure 8-2, one might conclude that all is well because actual costs are less than planned at this point in the project. Such a conclusion is totally unfounded because the project could be badly behind schedule and over budget if the true status of the work were known. Also, actual costs could exceed planned costs at a given point in the project because the project is ahead of schedule. In this case, exceeding planned costs at an intermediate point is not a bad sign.

The only point in a project when a comparison of actual versus planned cost is of value is at project completion. Unfortunately, many people use this comparison

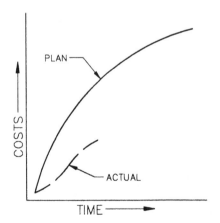

FIGURE 8.2.   Example cost comparison (the wrong way to track project progess).

as their only means of tracking a project in progress—a useless effort. The appropriate criterion is a comparison of the *value* of the work compared to the cost of the work to date. This also relates to the cost to complete the work compared to the remaining budget. This concept of "earned value," discussed in more detail later, is particularly effective because it integrates all project activities into a single value that can be accurately compared to plan.

There are other criteria for progress as well. Milestones completed or missed are a qualitative criteria for measuring how you're doing on the time scale. Usually, the end of a project task is very clear and easily compared to the task milestone. The units of work completed is a quantitative measure. For example, a construction project might involve pouring 1,000 yards of concrete over a one-month period. You could track progress as follows:

Day	Cumulative Progress Planned Yards	Actual Yards	% of Yards (Plan)	% of Yards (Actual)
5	100	80	10	8
10	250	200	25	20
20	700	710	70	71
30	1,000	1,000	100	100

This type of comparison demonstrated to the project personnel that added resources were needed, based on the results of the first 10 days. Appropriate adjustments were made between days 10 and 20 and the project ended on schedule.

Comparison with a project labor plan is also a useful way of tracking progress. You'll need to know the limits on productivity per person, the rate at which work can be done (i.e., only so many roofers can work on one roof at one time), and the relationship between tasks (doubling the number of drafters on a project isn't of value until the design group produces something to draw). Once these variables are known, you can track the actual labor versus plan and adjust resources as appropriate.

The earned value concept is the most effective criterion for tracking progress because it combines all facets of productivity into a single value. Also, it enables you to estimate how much it will cost to finish the project. The concept is best illustrated by an example. The example project involves producing ten computer runs over a four-week period (see Figure 8-3). The project plan is:

Project Element	Cumulative Time To Complete	Budgeted Costs	Cumulative Cost at Completion
Run 1	2 weeks	$2,700	$2,700
Runs 2 and 3	3 weeks	$900 each	$4,500
Run 4–10	4 weeks	$200 each	$5,900

The job is to produce
10 computer printouts

Total Computer Cost Estimate = $5,900

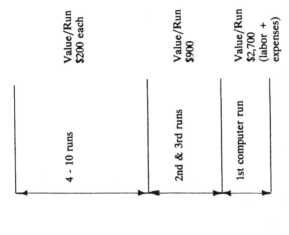

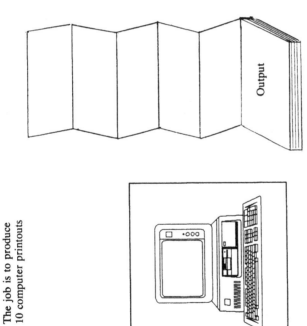

4 - 10 runs   Value/Run
                     $200 each

2nd & 3rd runs   Value/Run
                     $900

1st computer run   Value/Run
                     $2,700
                     (labor +
                     expenses)

10 runs represent all
of the Project Work
Packages

Output

FIGURE 8-3. Earned value concept.

The project plan is summarized graphically in Figure 8-4.

We will build a baseline budget for the job outlined under the following conditions:

Total project time estimated at four weeks.

Run 1 will be completed in two weeks. Runs 2 and 3 will be completed by the end of the third week. Runs 4 through 10 will be completed by the end of the fourth week.

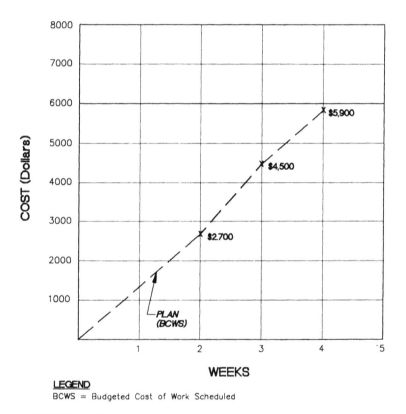

LEGEND
BCWS = Budgeted Cost of Work Scheduled

FIGURE 8-4. Earned value example—project plan

After three weeks, the project status is:

• At the end of the second week, Run 1 was completed at the estimated cost.
• Run 2 was completed, but Run 3 required more debugging.
• The program debugging required additional, unbudgeted costs. The cost-to-date

is $4,950. The debugging is now completed, and Run 3 can be completed in a straightforward manner in week 4.

Using the earned value criterion, you can determine project status, the potential cost overrun, and the potential schedule slippage. Plotting of the project data is shown on Figure 8-5.

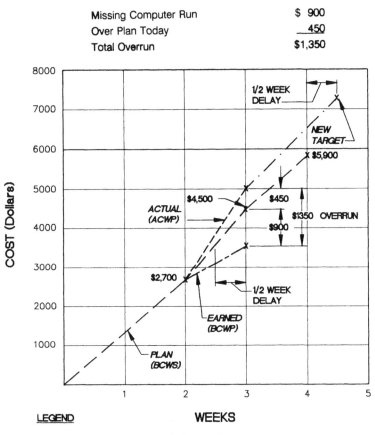

FIGURE 8-5. Earned value example—trend analysis.

The value of the work completed is $2,700 for Run 1 and $900 for Run 2, or a cumulative total of $3,600. The amount spent to date is $4,950. Therefore, the potential project overrun is:

Cost of Work to Date $4,950
Subtract Value of Work to Date   3,600
Overrun $1,350

The project status can be summarized in a simple graph, such as the one shown on Figure 8-6. Inspection of the graph after three weeks shows: 1) the actual cost of work completed exceeds the budgeted cost of work completed—the project is over budget; and 2) the time required is longer than the time planned—the project is behind schedule.

If the remainder of the work is completed at the rate projected in the initial plan, the project will be finished one-half week late. The revised project cost estimate and the new target then becomes $5,900 (original budget) plus $1,350 (overrun) or $7,250. The estimated cost to complete this is then $7,250 – $4,950, or $2,300. The earned value concept is relatively simple and extremely useful. It requires that each element or milestone within each element of the project be assigned a schedule and budget (value) at the start of the project. As each element or milestone is com-

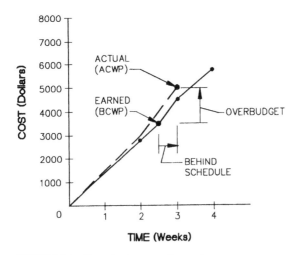

FIGURE 8-6.   Earned value example—project status.

pleted, the actual cost and time can readily be compared to the plan. Let's look at another example using production of a report with the following plan.

Project Element	Cumulative Time to Complete	Costs	Cumulative Cost at Completion
Research Literature	4 weeks	$10,000	$10,000
Prepare Draft Report	8 weeks	$10,000	$20,000
Receive Review Comments	10 weeks	-0-	$20,000
Prepare Final Report	14 weeks	$5,000	$25,000

The draft report was completed in nine weeks at a cost of $17,000. Comments on the report were received in 12 weeks. The project status is shown in Figure 8-7. The actual cost of work completed is less than budgeted, so the project is under budget. The actual time required is longer than that planned, so the project is behind schedule. In this case, assigning added resources to finalize the report may be appropriate if such resources could be productive (remember, only so many roofers can work on one roof at a time) and the schedule is critical.

Figure 8-8 illustrates project status graphs for a range of project conditions. In (a), the project is both ahead of schedule and under budget, and there is no need to make any corrections. In (b), the project is ahead of schedule and slightly over budget. The project is not nearly as far over budget as it would appear to be without

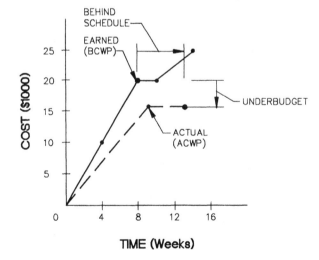

FIGURE 8-7.   Earned value example no. 2—project status.

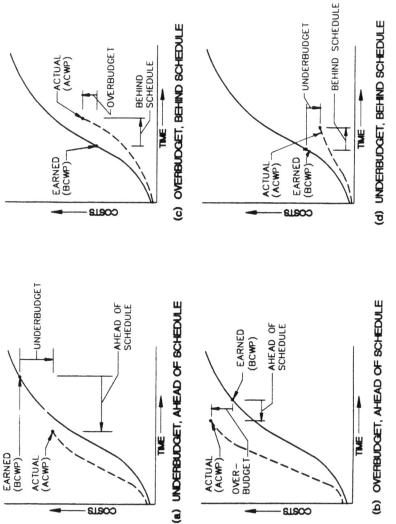

FIGURE 8-8.   Typical earned value status graphs.

consideration of the earned value. In this case (ahead of schedule, over budget), reducing the resources assigned to the project offers a potential method to reduce costs, depending on the nature of the project. By just comparing actual to budgeted costs, a hunt for scapegoats charging too much time to the project might have begun. In fact, the team has been very productive. In (c), the project is both behind schedule and over budget. A critical project review is needed to determine appropriate corrections. In (d), the project is under budget but behind schedule. Assigning added resources may be appropriate to make up time. By comparing actual to budgeted costs, no actions would have been taken when, actually, some was required to finish on time.

Simple examples were used to illustrate the earned value concept. On large projects with many tasks, computer programs are available to assist in the bookkeeping tasks associated with integrating the actual costs versus earned value for each task into a single report summarizing project status.

While it is important to know the overall project status at a given point in time, watching project trends can also be a very valuable monitoring tool. For example, if the labor charged to the project has been stable for several weeks but drops sharply below plan for two weeks, the project is headed for trouble, even if the overall project status is still on track. Watching trends may allow you to uncover a problem in time to resolve it before the overall project plan is irreversibly damaged. It would have been better in this case, to have been in communication with the team to know why the labor dropped before the cost data reached you.

Project Reviews

Be ever questioning. Ignorance is not bliss. It is oblivion. You don't go to Heaven if you die dumb.

Admiral Hyman Rickover

There is no such thing as the project that goes precisely according to plan. Project reviews are the critical tool you can use to catch deviations from plan and make adjustments before disaster occurs. You wouldn't want to fly in an airplane that doesn't receive frequent and thorough inspections. Likewise, you shouldn't expect to manage a project successfully without thoroughly reviewing its condition. You cannot rely on project team members to alert you to problems. They are only human. They are reluctant to bring bad news to the boss. People in the midst of doing project work tend to be overconfident about the time and cost to complete the work. They can lose perspective by focusing on only their part of the project. As a result, you will become insulated from project realities unless you periodically and personally review its progress. Reviews can run the gamut from informal exchanges with project personnel to formal reviews at scheduled points in the project. There is much to be learned from the informal exchanges you can get from the MBWA technique. Inherent in this technique is that you get out of your office and in touch with the project team.

You should include your internal formal review efforts in the project plan, as well as any external reviews with the client, and treat them as a small project within the main project. They should be carefully planned. On projects involving diverse technical disciplines, you may need to involve others in the reviews. Each person involved must know how much time and what detail of review they are expected to provide.

A key to reviews is that they be carried out in a friendly manner that is not threatening to the project team. You are very likely to hear some bad news, and there may be a natural tendency to lapse into blame and recrimination. Don't do it. Attack the problems, not the individuals (see Chapter 9 for more discussion on this point). If you attempt to gather information in an authoritarian (barging in whenever you please) or accusatory manner, you'll find doors slamming and information hard to get. By developing rapport, you'll have a much better chance of productive results.

Explain to the project team at the start of the project that reviews are an integral part of achieving a quality result. They are intended to discover what modifications of the overall project management system would better support the team. Point out the schedule and budget for any formal reviews that are scheduled. We've found, when properly explained, that the project team members will welcome the chance for some outside input to their work and to have a channel available to communicate their concerns. The reviews offer them feedback as to whether they are on course, as well as an opportunity to learn from the reviewer.

We need less emphasis on having all the answers and more on being sure we ask the right questions.

Donald Petersen

The best way to get information is to ask questions and *listen* (see Chapter 2 for basic communication skills that will be very useful). Some useful questions to have in mind include:

- Do you need any resources that you don't have?
- Do you know of anything that will cause you schedule or budget problems?
- Any chance you'll finish early? Under budget? Could you explain the reasons we're doing so well?
- What's the most difficult aspect of your work?
- What problems do you anticipate? What kinds of steps should we take?
- What's your assessment of the overall project status?

- What should we be doing differently?
- What percent complete is your work? What's the basis for your estimate?
- How much effort and time do you estimate that it will take to finish your work? What's the basis for your estimate?
- What problems are you having interfacing with other project tasks?
- Which incomplete tasks are in progress? Which ones haven't started?
- What other demands are there on your time?

If there is some tangible output (a task report, a draft project report, computer outputs, prototype construction, etc.), you should review it. Your own independent assessment of its status is important.

The frequency of contact with your team members must be based on good judgment. It is possible to over-control as well as to under-control a project. There have been cases where a manager's contact to discuss status creates a problem where none existed. The effort to control the project creates, in this case, a problem that then justifies the manager's attention. The extent and frequency of information gathering must be judged on a case-by-case basis.

Once you've gathered the information and made your evaluation of the project status, discuss your conclusions with the individuals on the project team before you make any modifications. By involving the team in reviewing your conclusions, they are more likely to accept them. Also, they are likely to have some good ideas on what may have gone wrong and how to correct the problems. The more the team is involved in uncovering and solving any problems, the greater the chance that the solutions will be accepted and effective.

Defining Project Changes

Follow the first law of holes: If you are in one, stop digging.

Dennis Healy

When you are getting off track, it's time to stop and do some analysis. There are a variety of causes to look for when a problem arises:

- Customer has changed their mind on schedule, cost, and scope.
- Inflation may have exceeded the amount allowed in the budget.
- People working on the project have changed.
- The initial plan was poor.
- Effort has been confused with real progress in assessing status.
- People are backtracking to make parallel work segments follow the same format or be compatible technically.
- Project tasks have been started and/or completed out of sequence.
- Customer requirements were not understood.
- Scope changes have occurred with no time or budget adjustments.

- In the desire for perfection, someone is putting in more effort than is justified.
- Project scope is ambiguous.
- Nonproductive people are working on the job.
- Costs have been erroneously charged to the project.
- Project reviews are inadequate or untimely.
- Scope has not been communicated well to team.

If your project doesn't work, look for the part you didn't think was important.

Arthur Bloch

Because it is impossible to anticipate every circumstance on a project, it is inevitable that changes will be necessary. Yet people have a natural reluctance to make formal changes in the project. Such an effort requires work to modify budgets and schedules and is an admission that the original plan was wrong. It is important that you establish a safe environment for people to make changes by focusing on what in the system allowed the problem to occur, as opposed to laying blame or seeking the "guilty."

Once a problem is uncovered, people on the team will feel psychological pressure to fix it as quickly as possible. After all, no one likes to tell the boss or customer about an unresolved problem on the project. However, a rush to solve the problem may miss both the real cause and the best solution. In the rush, an innocent bystander may become a sacrificial lamb, while a problem inherent to the organization or system is overlooked. A strong approach is to use the brainstorming and related techniques described in Chapter 9 to develop and evaluate several alternatives. Team involvement will increase acceptance of the changes.

It is impossible to generalize on solutions because each project is unique. There are some solutions with fairly wide application because they apply to common causes of problems. First, once you've defined the best option, implement it. Stop looking for something better. Stop throwing good money after bad. Consider getting the less experienced people off a project in serious trouble. The experienced members of the team will be more productive if they don't have to train others. If appropriate, consider using temporary help or giving some of the work to a subcontractor that you respect and trust—the added staff could get you back on schedule. Focus your resources on the critical or problem tasks until they are resolved. Consider having the team work overtime, keeping in mind that there is a point of diminishing return, where more overtime does not result in corresponding productivity. Data we've seen indicates that productivity will start to drop if a pace of more than nine hours/day is sustained over a long period of time.

Most executives lose their energy and productivity after about ten straight hours of work.

Business Month

Keep an eye on those who consistently work extra hours. Spending long hours at the office does not necessarily indicate an effective worker. More likely, it indicates a lack of personal effectiveness due to lack of needed tools or an inability to say no. It may indicate that the person only feels satisfaction through doing, doing, doing, and is fearful of offending others.

Consider involving the customer by suggesting to them ways that they could help expedite work, maybe by involving the customer's staff in some tasks. Check the project cost reports carefully—don't assume that they are accurate. Erroneous charges will obviously distort the cost picture, increasing the frequency of project milestones. Be sure that all project work is required in the scope and is not being done to an excessive level of detail (i.e., you asked what time it was and someone on the team built a watch).

The project schedule and budget will be impacted by a change, yet often no one bothers to actually change the schedule or budget. No one likes to admit that a change is needed or go through the hassle of altering the project plan. Another common reason is unfounded optimism that the change will have no effect, so there is no reason to change the plan. Even if the completion date isn't affected, individual task schedules and personnel assignments can be. Changes must be incorporated into the project plan, or any hope of control or accurate monitoring of progress is lost. Documentation of the change and its anticipated effect should be made when the change is initiated, not later. It becomes very difficult to sort out the effects after the fact. It is prudent to build the awareness of the project team on the importance of incorporating changes in the project plan at the start of the project.

If the scope and budget need to be changed, get experienced team members to come up with the revisions. Forcing revisions on the team is counter-productive because they won't buy into the changes without being involved in developing them.

There are some unplanned, unanticipated project changes that can change the course and outcome of the project. Be open to them. Don't let the project plan become a rigid impediment. This story, told by Walter Herby, illustrates the point. A lone shipwreck survivor on an uninhabited island managed to build a rude hut in which he placed all that he had saved from the sinking ship. He prayed for deliverance and anxiously scanned the horizon each day to hail any passing ship. One day he was horrified to find his hut in flames. All that he had was gone. To the man's limited vision, it was the worst that could happen. Yet the very next day, a ship arrived. "We saw your smoke signal," the captain said. Sometimes, positive results can come from unplanned project changes!

Status Reports

It is better to be brief than tedious.

William Shakespeare

Effective progress reports are one way to keep your customer and your management from distracting you with frequent inquiries as to what's happening on the project. Keep the reports as brief as possible, using pictures or simple graphics whenever possible. In deciding on a report format, put yourself in your customer's shoes. What are they really interested in knowing? On a $27,000,000 municipal construction project, we used a one-page monthly status report for the city council. The report consisted of an aerial photograph of the project taken at the first of the month, a graph presenting the earned value versus budgeted cost, and a one-half page list of any significant accomplishments or issues. The report was very effective. The photograph conveyed a tremendous amount of information in one glance, and was the level of detail needed for the council.

A common error is to cram too much detail into a status report. Often, upper management, yours and the customer's, want to quickly learn of the project status without all the details. They can ask for details if a key issue in your summary is of interest.

The frequency of reports depends on the duration of the project. Obviously, reports can be much less frequent on a ten-year duration project (quarterly will probably do), as compared to a ten-month project (monthly will be needed).

Project Completion

The Six Phases of a Project . . .
1. *Enthusiasm*
2. *Disillusionment*
3. *Panic*
4. *Search for the Guilty*
5. *Punishment of the Innocent*
6. *Praise and Honor for the Nonparticipants*

Anonymous

The above illustrates the fact that the euphoria that accompanies the start of a project usually vanishes as the project nears completion. Project team members are already looking forward to their next assignment and may not pay attention to the details of project completion. Those with no new assignment in sight are in no hurry to finish and may stretch out the project. Effective project completion can be the difference between success and failure because project profits can be dissipated, the customer can become upset, and the transfer of technical outputs is delayed, reducing the learning that others could achieve from the project. Your customer will better remember the most recent events, so the way you close out the project can have a substantial effect on your chances for follow-on work.

Most project managers have a relatively easy time understanding the technical and mental chores of finalizing project reports, assembling the final cost information, finishing any tasks, and disposing of any project hardware. They are not as comfortable dealing with the emotional issues that arise as project completion approaches. To address these issues, you must consider and deal with the needs and personality of each team member on an individual basis. The following emotional issues are typically encountered:

- **Loss of interest**—As the job nears completion, the creative aspects are replaced by more mundane administrative duties. Creative people become bored, disenchanted, and look upon the remaining work as drudgery. If not removed from the project, they may get little done because they will divert their energy to more interesting tasks or in seeking another assignment. Pick someone who is good at details and patient and methodical with paperwork.
- **Anxiety over the future**—If there has been no specific future assignment, a person's anxiety may lead to foot-dragging. The slowdown may result from a conscious decision to spend time looking for another job or from a subconscious desire to stretch out the remaining work, to provide some security. If the next assignment is known, there will be anxiety if it is viewed as a less attractive assignment.
- **Loss of motivation**—In the height of an exciting project, team members are bound together by a common mission that motivates them. As the project nears completion, team objectives become less important and individual concerns take over. The remaining work is unexciting details, and motivation drops.
- **Lack of focus**—Work on more interesting new assignments can disrupt the focus of the project personnel. Their energy soon becomes scattered between the old and new projects, with the old one losing out.
- **Pressure to remain billable**—If the project hasn't been carefully planned, the funds to close the job may be marginal or insufficient. Team members will feel pressured to work on other projects, where funds are available.

Do not turn back when you are just at the goal.

Publilius Syrus

To effectively close out the project, consider restructuring the project team and treating the project closeout as a separate mini-project, with its own plan, budget, and schedule. You'll need to focus on the emotional issues of the team members before you can proceed with an effective project completion. Consider the personality and skills of each team member. Retain the personnel who deal well with details and who understand the project scope and contract. Proceed immediately to reassign the more creative members of the team to other projects. Your positive actions in these reassignments and in restructuring the remaining team members will calm anxieties and reduce the energy that would be lost to discussing rumors about the future. Even if reassignment is not possible for everyone on the team, an honest, open approach is much more positive than inaction that leaves everyone on edge. Treating all team members honestly and openly by discussing information about their next assignments demonstrates your respect for the individuals on your team and will assist in motivating those that remain to close out the project.

To motivate the remaining team, define the objectives of the close-out effort as though it were a project in itself. Get the team involved in defining these objectives, setting a schedule, and dividing up assignments. Keep communications going by informal staff meetings and by frequent, individual contact with each team member during the close-out phase. It is vital to personally demonstrate your sense of urgency and enthusiasm. Be generous with pats on the back, recognizing the unrewarding nature of this phase of the project. By taking the same pride in completing the project as you did in executing the project, you'll set an example for the team to emulate. The time required will be condensed so that you can all move on to the next project, taking pride in having efficiently completed the last project. Have a ceremony or party at the completion of the project to: 1) provide recognition; 2) acknowledge team effort and spirit; 3) provide a sense of completion; and 4) set a precedent that will give future teams something to look forward to.

Post Evaluations

The only thing new in the world is the history you don't know.

Harry Truman

The way that you and your organization can learn is by observing the consequences of your actions on the project, make inferences about these consequences, and make conclusions about how to act on future projects. Most people don't do this. Maybe it's too painful. Maybe it's threatening to the organization. We suggest that following project closeout, you reassemble the team for a freewheeling discussion of project performance. Make it a post-project success acknowledgement as well. With the right attitude, this session can be fun. Brainstorm (see Chapter 9 for brainstorming discussion) questions such as:

- How did our final costs and scope compare with the project plan?
- How did our end-result compare with the project plan?
- How was our communication with our customer? How did it change as the project progressed?
- How was our communication among team members?
- How did we perform as a team and as individuals?

- What would we do differently next time?
- What would we do the same next time?

By documenting the results of this discussion, you provide a valuable resource to be used by teams that follow, as well as for yourself. You'll gain insights into scoping future projects, dealing with a client's wants and needs, scheduling tasks, and where time was lost or gained. The gains to be had are great in comparison to the time and effort involved.

Experience is not what happens to a man. It is what a man does with what happens to him.

Aldous Huxley

References

1. Truman, J. 1988. Development and implementation of project management systems. *Project Management Handbook,* edited by D. Cleland and W.R. King, p. 652. New York: Van Nostrand Reinhold.
2. Bent, J.A. 1988. *Project Control: An Introduction,* edited by D. Cleland and W.R. King, p. 559. New York: Van Nostrand Reinhold.
3. Chilstrom, K.O. 1988. *Project Needs and Techniques for Management Audits,* edited by D. Cleland and W.R. King, p. 620. New York: Van Nostrand Reinhold.
4. Levine, H.A. 1988. *Computers in Project Management,* edited by D. Cleland and W.R. King, p. 692. New York: Van Nostrand Reinhold.
5. Rosenau, M.D., Jr. 1984. *Project Management for Engineers.* New York: Van Nostrand Reinhold.
6. Hamburger, D.H., and H.G. Spirer. 1989. Project completion. *Project Management, A Reference for Professionals,* edited by R.L. Kimmons and J.H. Loweree, p. 587. New York: Marcel Dekker, Inc.
7. Horowitz, M.E. 1989. Monitoring progress. *Project Management, A Reference for Professionals,* edited by R.L. Kimmons and J.H. Loweree. p. 543. New York: Marcel Dekker, Inc.
8. Cox, W.R. 1989. Control of changes. *Project Management, A Reference for Professionals,* edited by R.L. Kimmons and J.H. Loweree. p. 561. New York: Marcel Dekker, Inc.
9. Kerzner, H. 1989. *Project Management, A Systems Approach to Planning, Scheduling and Controlling.* New York: Van Nostrand Reinhold.
10. Zeldman, M. 1989. "Project Planning, Scheduling and Control," unpublished seminar materials. Pittsburgh, Pennsylvania: EmZee Associates.
11. Burstein, D., and F. Staslowski. 1982. *Project Management for the Design Professional.* New York: Whitney Library of Design.
12. Graham, R.J. 1989. *Project Management As If People Mattered.* Bala Cynwyd, Pennsylvania: Primavera Press.

9

Quality Management

Get the confidence of the public and you will have no difficulty getting their patronage . . . Remember always that the recollection of quality remains long after the price is forgotten.

H. George Selfridge

A quality result is the ultimate goal of a project. The skills and procedures in the earlier chapters must be integrated to produce this result. The value of quality has long been recognized. The Code of Hammurabi, dating from 2150 B.C., states, "If a Builder has built a house for a man, and his work is not strong, and the house falls in and kills the householder, that the builder shall be slain." Phoenician inspectors eliminated any repeated violations of quality standards by chopping off the hands of the makers of defective products.

What is relatively new is the application of formalized techniques developed by Dr. W. Edwards Deming to service industries. Dr. W. Edwards Deming is often considered the father of modern quality management, referred to as "Total Quality Management" (TQM). Dr. Deming took his concepts to Japan after World War II and was largely responsible for the amazing improvement in the quality of Japanese products. "Made in Japan" went from being a synonym for junk to being a synonym for the highest quality products as a result of a process that began in the early 1950s. One of the greatest levels of recognition in Japan is to receive the Deming Award for Quality.

In 1980, NBC aired a white paper program, "If Japan Can, Why Can't We?" The program elevated the awareness of the American manufacturing industries to TQM concepts and techniques. Ford Motor Company, among others, began to apply the techniques. In 1987, President Reagan signed a law that created a national award for quality for up to six companies per year—the Malcolm Baldrige

National Quality Award. Two awards per year can be made in the categories of manufacturing, service companies, and small businesses. Motorola, Xerox, Cadillac, IBM, and Federal Express are among the winners to date. The award focuses strongly on customer satisfaction. The guidelines for the award make an excellent self-examination guide for quality and are presented in Table 9-1. The application, even if it is never submitted, is such an in-depth process that it is an excellent tool to use in evaluating your organization.

These guidelines change from year to year, so the detailed scoring criteria may change. However, the basic framework remains the same. The types of things evaluated in the seven categories provide insight into the elements of a successful program to manage quality:

- **Leadership**—"How senior executives create and sustain clear and visible quality values along with a management system to guide all activities of the company toward excellence."
- **Information and analysis**—"The scope, validity, use, and management of data and information that underlie the company's overall management system."
- **Strategic quality planning**—"The company's planning process for achieving or retaining quality leadership, and how the company integrates quality improvement planning into overall business planning."
- **Human resource utilization**—"The effectiveness of the company's effort to develop and realize the full potential of the work force—including management—and to maintain an environment conducive to full participation, quality leadership, and personal and organizational growth."
- **Quality assurance of products and services**—"The systematic approaches used by the company for assuring quality goods and services based primarily upon process design and control, including control of procured materials, parts and services."
- **Quality results**—"The company's knowledge of the customer, the overall customer service systems, responsiveness, and its ability to meet requirements and expectations. Also examined are current levels and trends in customer satisfaction."
- **Customer satisfaction**—"The quality levels and quality improvement based upon objective measures derived from analysis of customer requirements and expectations, and from analysis of business operations. Also examined are current quality levels in relation to those of competing firms."

If you desire more information on the award and related procedures, you can obtain it from the National Institute of Standards and Technology, Gaithersburg, Maryland (301-975-2036).

The commercial value of recognition for quality was evidenced in 1990 when Cadillac advertisements began featuring their award on the *same day* that notice of the award was made public. The increased interest in quality management is

TABLE 9-1   Malcolm Baldrige National Quality Award
Examination Categories

1990	Examination Categories/Items	Maximum	Points
1.0	**Leadership**		**100**
	1.1 Senior Executive Leadership	30	
	1.2 Quality Values	20	
	1.3 Management for Quality	30	
	1.4 Public Responsibility	20	
2.0	**Information and Analysis**		**60**
	2.1 Scope and Management of Quality Data and Information	35	
	2.2 Analysis of Quality Data and Information	25	
3.0	**Strategic Quality Planning**		**90**
	3.1 Strategic Quality Planning Process	40	
	3.2 Quality Leadership Indicators in Planning	25	
	3.3 Quality Priorities	25	
4.0	**Human Resource Utilization**		**150**
	4.1 Human Resource Management	30	
	4.2 Employee Involvement	40	
	4.3 Quality Education and Training	40	
	4.4 Employee Recognition and Performance Measurement	20	
	4.5 Employee Well-Being and Morale	20	
5.0	**Quality Assurance of Products and Services**		**150**
	5.1 Design and Introduction of Quality Products and Services	30	
	5.2 Process and Quality Control	25	
	5.3 Continuous Improvement of Processes, Products and Services	25	
	5.4 Quality Assessment	15	
	5.5 Documentation	10	
	5.6 Quality Assurance, Quality Assessment and Quality Improvement of Support Services and Business Processes	25	
	5.7 Quality Assurance, Quality Assessment and Quality Improvement of Supplies	20	
6.0	**Quality Results**		**150**
	6.1 Quality of Products and Services	50	
	6.2 Comparison of Quality Results	35	
	6.3 Business Process, Operational and Support Service Quality Improvement	35	
	6.4 Supplier Quality Improvement	30	
7.0	**Customer Satisfaction**		**300**
	7.1 Knowledge of Customer Requirements and Expectations	50	
	7.2 Customer Relationship Management	30	
	7.3 Customer Service Standards	20	
	7.4 Commitment to Customers	20	
	7.5 Complaint Resolution for Quality Improvement	30	
	7.6 Customer Satisfaction Determination	50	
	7.7 Customer Satisfaction Results	50	
	7.8 Customer Satisfaction Comparison	50	
	TOTAL POINTS		**1000**

shown by the change in the number of requests for the Baldrige criteria, often used by companies to assess their own quality programs. In 1988, 12,000 copies were requested. Over 180,000 applications were requested in 1990, with 1,200 submitted. Xerox reportedly spent $800,000 pursuing the award (successfully) in 1989.

In the late 1980s, service industries (such as insurance companies, Federal Express, and consulting engineers) began to apply the concepts and tools of TQM, and application techniques are still evolving. This chapter will look at potential applications of TQM to project management and the integration of the basic tools described in Chapters 1 through 8 into quality management. We will introduce the basic concepts here. The reader is encouraged to use the references cited for more details.

WHAT IS QUALITY?

Our understanding of quality is still evolving. A basic concept of quality is conformance to a standard with minimum variation. This requires design of standards for services and products and parameters to measure variations from the standard. But "conformance to a standard" only provides a primitive understanding of quality. The design of the standard to fit the needs of the client and the continual improvement of the product and service are crucial aspects of quality.

Although services are an intangible, service quality is anything but an intangible concept to the people who use them.

Charles Lee
President and CEO, GTE Corporation

The concept of "quality" in a service industry requires measuring subjective values. In manufacturing, it is simple to visualize objective measures of quality. The final product is a tangible item that can be weighed and dimensioned and compared to specifications. Manufacturing quality can then be quantified by how closely the product conforms to the specification. For service industries, objective specifications are elusive for criteria such as reliability, responsiveness, accuracy of opinions, and courtesy.

For a project, you might think that quality has been achieved if you complete the job on time, within budget, and meet all technical specifications. But quality is more than these three ingredients. For example, Baker et al. researched the results of 650 completed projects, to determine what constituted success or quality. They found a surprising array of factors that described project failure or success, as shown in Table 9-2.

They then defined a quality project:

If the project meets the technical performance specifications and/or mission to be performed, and if there is a high level of satisfaction concerning the project outcome

TABLE 9-2   Project Success Factors (Reference 5)

Successful Project Characteristics

- Frequent feedback from the parent organization.
- Frequent feedback fro the client.
- Judicious use of networking techniques.
- Availability of back-up strategies.
- Organization structure suited to the project team.
- Adequate control procedures, especially for dealing with changes.
- Project team participation in determining schedules and budgets.
- Flexible parent organization.
- Parent organization commitment to established schedules.
- Parent organization enthusiasm.
- Parent organization commitment to established budget.
- Parent organization commitment to technical performance goals.
- Parent organization desire to build up internal capabilities.
- Project manager commitment to established schedules.
- Project manager commitment to established budget.
- Project manager commitment to technical performance goals.
- Client commitment to established schedules.
- Client commitment to established budget.
- Client commitment to technical performance goals.
- Enthusiastic public support.
- Lack of legal encumbrances.
- Lack of excessive government red tape.
- Minimized number of public/government agencies involved.

Unsuccessful Project Characteristics

- Insufficient use of status/progress reports.
- Use of superficial status/progress reports.
- Inadequate project manager administrative skills.
- Inadequate project manager human skills.
- Inadequate project manager technical skills.
- Insufficient project manager influence.
- Insufficient project manager authority.
- Insufficient client influence.
- Poor coordination with client.
- Lack of rapport with client.
- Client disinterest in budget criteria.
- Lack of project team participation in decision making.
- Lack of project team participation in major problem solving.
- Excessive structuring within the project team.
- Job insecurity within the project team.
- Lack of team spirit and sense of mission within project team.
- Parent organization stable, non-dynamic, lacking strategic change.
- Poor coordination with parent organization.
- Lack of rapport with parent organization.
- Poor relations with parent organization.
- New "type" of project.

Table 9-2   Project Success Factors (Continued)

Unsuccessful Project Characteristics

- Project more complex than the parent organization has completed previously.
- Initial under-funding.
- Inability to freeze design early.
- Inability to close out the effort.
- Unrealistic project schedules.
- Inadequate change procedures.
- Poor relations with public officials.
- Unfavorable public opinion.

among key people in the parent organization, key people in the client organization, key people on the project team, and key users or clientele of the project effort, the project is considered an overall success.

Quality is the result of doing the right thing and doing it right the first time, on time, so as to meet the customer's expectations and needs.

QUALITY MANAGEMENT VERSUS QUALITY CONTROL

The term "quality control" is used to describe inspection or checking of work products. In simple terms, quality management is designed to prevent defects by doing the job right. Management is concerned with preventing problems by creating the attitudes and controls that make prevention possible. The concepts of person-centered project management are a good foundation. The skills, practices, and attitudes described in the first eight chapters of this book can all be applied in achieving a quality management program. Some of the key elements of quality management are:

- Continuous process improvement.
- Total employee involvement.
- Prevention, not correction.
- Supplier-customer relationship.
- Customer satisfaction.
- Problem-solving teams.
- Tools.
- Measurement.
- Recognition.

BASIC CONCEPTS

Deming's Fourteen Points

Dr. Deming has summarized his theories in his famous "Fourteen Points":

1. Create consistency of purpose toward improvement of product and service, with the aim to become competitive, stay in business, and provide jobs.
2. Adopt the new philosophy. We are in a new economic age. Management must awaken to the challenge, must learn responsibilities, and take on leadership for change.
3. Cease dependence on inspection to achieve quality. Eliminate the need for inspection on a mass basis by building quality into the product in the first place.
4. End the practice of awarding business on the basis of price tag. Instead, minimize total cost. Move toward a single supplier for any one item on a long-term relationship of loyalty and trust.
5. Improve constantly and forever the system of production and service, to improve quality and productivity, and thus constantly decrease costs.
6. Institute training on the job.
7. Institute leadership. The aim of leadership should be to help people, machines and gadgets to do a better job. Supervision of management is in need of overhaul, as well as supervision of production workers.
8. Drive out fear, so that everyone may work effectively for the company.
9. Break down barriers between departments. People in research, design, sales and production must work as a team to foresee problems of production and in use that may be encountered with the product or service.
10. Eliminate slogans, exhortations, and targets for the work force that ask for zero defects and new levels of productivity without providing methods.
11. Eliminate work standards (quotas) on the factory floor. Substitute leadership. Eliminate management by objective. Eliminate management by numbers, numerical goals, substitute leadership.
12. Remove barriers that rob the hourly worker of his right to pride of workmanship. The responsibility of supervisors must be changed from stressing sheer numbers to quality. Remove barriers that rob people in management and engineering of their right to pride of workmanship. This means abolishment of the annual merit rating system.
13. Institute a vigorous program of education, re-education and self improvement.
14. Put everybody in the organization to work to accomplish the transformation. The transformation is everybody's job.

Deming's book, *Out of Crisis*, presents a full discussion. The basic, underlying concepts fall into the categories of:

- Focus on the customer needs.
- Employee involvement.
- Remove barriers.
- System concepts.
- Statistical process control.
- Minimizing variation from the norm.
- Training.
- Leadership.
  The balance of this chapter discusses these topics.

Focus on the Customer

Every company's greatest assets are its customers, because without customers, there is no company.

Michael LeBoeuf

Who can put a price on a satisfied customer, and who can figure the cost of a dissatisfied customer.

Dr. W. Edwards Deming

The requirements by which quality is judged are those of the customer. After all, customer satisfaction is the ultimate goal of an organization and any project team. You can very efficiently produce a beautiful product but find yourself out of business if it does not meet the customer's needs. Have you ever delivered a report or other product of your work to a client, only to be shocked to find it wasn't what he or she wanted? If so, your understanding of the client's needs obviously fell short. The customer's needs must have top priority if you are to succeed.

One of the classic American examples of not meeting the client's needs is the Edsel automobile, introduced by the Ford Motor Company in the 1950s. "Edsel" has become a commonly used label for the phenomenon of producing a product that customers don't want. The project team that worked on the Edsel, in fact, incorporated some advanced features for a car in its price range at the time. The entire Ford Motor Company was convinced that it was introducing a high-quality product. Unfortunately, it didn't meet the customers' needs.

This example illustrates the dual nature of quality—one that has been called "Quality in Fact" and "Quality in Perception." The provider of services or goods that meet their own specifications achieves Quality in Fact. A service or product that meets the customer's expectations achieves Quality in Perception. The Edsel met all of Ford's specifications achieved Quality in Fact but not Quality in Perception. For sustained success, you must pay close attention to both aspects of quality. A recent example is "New Coke," developed by Coca-Cola Company. The product met the company's own demanding specifications, but the customers were not impressed. Coca-Cola ended up with a large inventory of a quality product (i.e., one that met specs) that couldn't be sold. Low perceived quality can be a disaster.

Quality is what the customer says it is. Period.

Ron Zemke

Obviously, communication about the customer's needs at the start of a project is essential. The best way to learn about your customer's needs is to *listen* (see Chapter 2). If you're dealing with one specific customer, you can ask him or her some open-ended questions about his or her needs, such as "What do you see as the most critical elements of the project?" or "What would a successful project look like to you?" Be careful not to project your own preconceptions onto the customer about what constitutes a successful project. You can protect against this by checking your perceptions with questions like, "My understanding is that minimum operation and maintenance cost is the most critical element to you; is that correct?" If you're analyzing the needs of a group of customers, use market research to analyze their needs. Taking time up front to define these needs can minimize future delays and problems. The emphasis clearly needs to be on meeting these needs, as opposed to any intermediate or internal result.

. . . there's one thing no business has enough of: customers. If you want to have them, what you do after *the sale matters more than what you do to get it.*

Harvey MacKay

Focusing on the customer will produce attitude changes. We've seen cases where someone gets frustrated and upset when a call from a customer interrupts their work on a proposal or an internal project. By constantly focusing on serving the needs of your customers as your top priority, you see that you are performing your work best when you serve the needs of the customer. When the customer has a problem, focus on solving the problem and not on who in your organization may have contributed to the problem. This shift in attitude will increase the potential to find in others knowledge or resources that will help the customer. Committing to the service of the customer raises you above provincial interests and frees you from protecting your own job or your department's turf. It lifts you from the struggle to win credit for yourself or discredit others. It simply encourages teamwork.

Everyone in an organization has suppliers (people up the line) and customers (people down the line). Within your own organization, every step of the project involves internal suppliers and customers. Understanding their needs is important. For example, an employee wants a computer and to do so must use the purchase order process through the corporate purchasing department. The staff member is the supplier and the purchasing agent is the customer in the first step in this process. The staff member must understand what information the purchasing agent requires to purchase the system. If the staff member doesn't, there may be several inefficient interactions between the two people before the right purchase order is issued. In the next step, the purchasing agent becomes the supplier and must understand the information needs of the vendor, to insure that the right system is delivered. For a project team to meet the needs of their external customer, each team member has to meet the needs of his or her own internal customers (i.e., other team members). Many team members seem reluctant to ask their internal customers for feedback. They are not used to asking fellow workers questions like "What can I do to make you more productive?" "Am I meeting your needs?" and "What can I do better?" It often takes some training and practice in a safe environment to get these internal feedback loops to function.

In service industries, where the products are ideas, it is a challenge for everyone in the organization to relate responsibility to the final customer. One key to getting everyone up and down the line in a service industry to relate to attaining quality is to give every employee a well-defined customer. For many large service organizations, focusing on the needs of an internal customer is the answer because a relatively small percentage of the staff will have direct interface with the clients. The customer is anyone to whom you provide information, service, or a work product. The ultimate customer is outside of your organization, but the customer that a secretary or technician can do something about and get feedback from may be in an office a few feet away. You can ask your internal customer how you are doing.

Thinking of others in your organization as customers has several benefits:

- Because many people never see the external customer, it is hard for them to visualize the impact that their work has on the customer. When they look at a coworker who uses their output as a customer, it increases their awareness of customer satisfaction.
- When the person down the line from them is viewed as a customer, peer pressure for top-quality work increases.
- If you can demand top-quality work from those upstream of you, you know that those downstream from you have a right to top-quality work.
- Everyone on your team becomes more aware of quality, and problems get resolved more quickly and effectively.

Another key is to get everyone to understand how their function relates to the *core* function of the company. Good communication and teamwork are needed at every step, internally and externally. As internal suppliers and customers begin to communicate, surprisingly immediate quality improvements can result. Someone may be doing a task or providing a level of precision that really isn't needed. In one of the examples cited by Townsend for the Paul Revere Company, a clerk followed instructions from her predecessor and spent time each week sorting a stack of data cards in alphabetical order. When she began to communicate with the person she gave the cards to, she found that the cards were simply being thrown away. The need to review the cards had disappeared, but this information was never communicated. The card sorter was producing a quality job of sorted cards, but by not meeting a need of her internal customer, she was failing the test of quality. If each person in the internal chain of production treats the next person as a valued customer, quality in fact and in perception will result.

Consumer research can and should be performed internally as well as externally. Employees are, for example, customers of management policy decisions. Employee surveys can lead to improved policies. Field offices are often customers of home office services in large, multi-office firms. The home office may well learn how to improve quality by surveying the field offices, considering them as customers. Gathering information on internal or external customer needs is not a one-time project, but rather an ongoing, continual process. Some firms have adopted, as part of a quality improvement program, a requirement that each department manager interview at least one client per month, to obtain feedback on the firm's performance.

Employee Involvement

Only the people doing the work know how to improve it.

Edward Baker
Ford Motor Company

There is much in common between TQM concepts and concepts for person-centered management, as described in this book. It is critical to recognize that employees at all levels have good ideas and that managers can benefit by using them. An organization performs best when all members have a chance to contribute their ideas and share in the decision making. People prefer to work when their thoughts are heard and considered. The concept is to put everyone in the organization on a team. The teams are made up of different levels of the organization, a vertical slice through the organization chart rather than a horizontal one. Often, managers simply don't have all the information needed to make a good decision. Professional football quarterback Francis Tarkenton let his linemen suggest plays because, he says, "They were closer to the action than the head coaches pacing up and down the sidelines." Involvement in identifying and solving problems increases individual commitment and performance.

Jack Shewmaker, former president of Wal-Mart Stores, says, "Time after time, I have seen struggling businesses where the solutions were known to employees." But nobody asked. Examples abound, as companies have asked their staff for suggestions as part of an overall TQM program. Lockheed Corporation credits a $3,000,000 savings in its quality improvement program to small task groups. Johnsonville Foods, a sausage manufacturer, achieved a 20-fold growth in six years after turning over nearly all management functions, including hiring and firing, to the employees. A secretary's idea on beginning a catalog sales program for their food program grew to a one million dollar per year business in one year. Harley-Davidson, the manufacturer of motorcycles, is a well-known story of a company saved from bankruptcy by trusting the employees to restructure the production process and organization—27 manager positions were reduced to 1 and quality improved! General Motors turned over management of a floundering Bay City, Michigan production facility to a psychologist who instituted a program to make plant employees true partners in the production process. Output increased 40 percent, and rejects decreased 40 percent. Honeywell used its quality control circles to reduce the cost of one of its electronics products by 20 percent and subsequently won a major contract. American Airlines "Ideas in Action" campaign saved $41 million in 1988, based on employee suggestions. For example, a flight attendant suggested reusing plastic covers on coffee pitchers, saving $60,000.

Chaparral Steel in Texas offers another example. The average American steelworker produces 350 tons per year. The Japanese steelworker produces 850 tons per year. The Chaparral steelworker produces 1,350 tons per year. At Chaparral, the employees do all the hiring and firing, run the quality control program, run the safety program, and make the day-to-day decisions on the mill floor. Workers have easy access to the top manager because excess layers of management have been stripped away.

Cadillac Motor Car Division of General Motors began a program in 1987 that gave the people who assemble the cars input into the design. As a result, reliability problems dropped by 71 percent in the next three years. Customer satisfaction increased 16 percent.

Sure, you may be saying, these are manufacturers. Show me a service industry example! Townsend and Aebhardt have described the success at the Paul Revere Insurance Company. Their quality management program required that *all* of the company's 2,400 employees be active members of a quality team. In the first year of the program, the company implemented 4,110 ideas generated by the teams and saved $8,500,000. By the spring of 1988, over 26,000 ideas with a savings of over $16,000,000 had been implemented. The company's fall to second place in market share in 1983 was the catalyst for the program. Since 1985, the company has led in market share in its primary product, and the margin is widening. Employee involvement works in service industries!

Giving employees the information and the freedom to act is a powerful tool in obtaining the commitment needed to assure quality work. If you're finding, for example, that the hours required to produce each sheet of drawings in a design project are above the norm for your projects, disseminate the information to your project team. Solicit input and suggestions from the engineers, designers, and draftspeople. Get the entire team involved in identifying problems and solutions.

Such involvement will often lead to surprising creativity. Creativity is usually just elaborating on current methods, putting things together in a different way, or taking something away from the present to create something simpler or better. Large doses of incremental improvements can make a company far more effective than a one-time attempt at innovation or change. Anyone in a company has the capacity to contribute to innovation. People expect more from themselves when they have a chance to go beyond the routine and contribute.

Remove Barriers

A man can succeed at almost anything for which he has unlimited enthusiasm.

Charles Schwab

Most organizations inherently, although unknowingly, are replete with barriers to pride of workmanship, quality, and enthusiasm. Every rational manager would certainly say that he or she is in favor of quality, but, at the same time, he or she may well be oblivious to his or her own acts and policies that inhibit quality work.

Barriers between groups within a company are often a major hindrance to the teamwork essential to achieving quality work—marketing versus production, production versus accounting, field people versus office people, and so on. To break down these barriers, you must define how everyone's efforts contribute to meeting the customer's needs. This can create a change in attitude so that the employees relate to and cooperatively pursue a unifying organizational goal rather than their departmental or personal goals. This attitude shift is crucial in getting effective work sharing between offices or departments, when one department has the resources needed by another to meet the customer's needs. Without the shift, each office will focus on its own bottom line, to the detriment of the customer and, in the long run, to the organization.

A quality environment encourages teamwork, communication, joint problem solving, trust, security, and pride of workmanship. Teamwork is essential to foster continual improvement and quality work. It will also improve your competitive position. The sports world offers many examples where a team with superior individuals loses to another group who are truly functioning as a team. A winning coach creates a game plan that allows each team member to contribute so that the resulting team effort is greater than the sum of the individuals' talents.

The manager's job is to create an environment without barriers so that individuals' talents come together synergistically to produce a team result. Management to achieve quality is much like that of the winning coach or like that of an orchestra leader who creates beautiful music from a group of talented individuals who, without the leader, would not have produced the same result.

Deming and other TQM researchers have concluded that the annual performance appraisal system used so commonly in the United Sates is a major barrier to quality work. You may have noted in the earlier listing of Deming's fourteen points (see page 208, point number 12) that they include abolishing the annual rating system. You can see that removing barriers may involve doing damage to some sacred cows.

One of the problems with the appraisal system is that the boss becomes the employee's most important customer. After all, it is the boss who fills out the annual appraisal form. Unless you get a good rating from the boss, the security of your family and career will be threatened. As a result, the true customers (the ones who use his or her output) suffer. Also, teamwork is damaged because your performance rating depends on meeting your individual goals, not on the team or others meeting theirs. Since all employees have their own goals and related ratings to worry about, they are all simultaneously pursuing different and possibly conflicting goals, with the likelihood that the overall company results will suffer. For example, consider the office manager who is trying to reduce overhead costs and the office marketer who has a goal to increase new project fees by 20 percent. The added overhead costs for the greater marketing effort will include elements that are in direct conflict with the office manager's goals (added secretarial labor, more staff engineering time spent on proposals, added expenses, etc.). It is likely that the two will be working at cross-purposes at least part of the time.

The performance system also encourages mediocrity because it reduces risk-taking. Usually, your annual goals are set in conjunction with your boss. Typically, you'll do all you can to make sure that the goals minimize the risk of failure. Most bosses go along with the process of achievable goals because they like to see people feel like winners. This process can also lead to a lot of "sandbagging" and results that are less than what could have been produced. Consider yourself as the marketer who had the goal to increase fees by 20 percent. As the year nears the end, you see that you will achieve the 20-percent goal in November. You delay other work that could have been signed up before year-end so that you can bankroll it for next year. As a result, the company does not really know the full potential of the market. One of the prime examples of this technique was Vasili Alexeyev, a Russian weightlifter. He was given a reward for each world record that he broke. He quickly figured out that he wanted to break a lot of records, but only by a gram or two at a time so that he could keep breaking his own records. People quickly learn how to work the system!

Performance appraisals increase emphasis on the short term, often at the detriment of long-term results. The department manager can meet his goal of reduced overhead by eliminating marketing costs for the next six months. This will look great on the manager's performance rating this year, but the company will pay for it in terms of fewer new projects in the future. A reasonable long-term view of the balance between profit and quality is essential to quality work. A typical performance criteria for project managers is meeting project schedules. As a result, highly visible project milestones may be met by getting the required products submitted, regardless of their completeness or quality. The manager gets a good rating for meeting schedules, and the company later pays the price to rework the premature submittals.

No matter how quantitative performance goals may appear, judgment on performance still tends to be subjective. Ratings by different managers within the same company are usually not comparable. Performance earning an excellent rating from one manager may receive only a good rating from another. As a result, there is not a consistent pattern for quality results that can be emulated by all. Delay in feedback until a periodic performance review creates frustration when good performance is not quickly recognized and anger when judgments are made about problems that are now history.

Appraisals are often used to determine salary increases, yet do not provide enough information about the subtleties of performance. As a result, differentials in salary adjustments are arbitrary and bear no relationship to the quality of work.

I am like an ant on the front of a log heading downstream toward a treacherous bend and all I can do is stick my foot in the water to try to steer us clear and yell, "whoa, you SOB, whoa."

R.C. Gerstenbery
Former Chairman, General Motors

The annual review process assumes that each person is directly responsible for his or her output. As discussed in "Systems Concepts" (see page 219), this assumption overlooks the effect that the *system* has on each individual's results. This shifts the blame for system problems that are management's responsibility to the employees, and it hits them right in their pocketbooks when their appraisals suffer. Morale and pride in their work suffer.

Another major problem results from the way that performance systems are typically administered. It is typically done in a manner worse than amateur night for aspiring comedians. Most supervisors are not trained to effectively communicate with subordinates about their strengths and weaknesses, and many are not eager to do so. Often, they look upon the review process as another administrative burden to be dispensed with as quickly and painlessly as possible. The forms are ritually completed and filed, with neither the company nor the employee profiting from the experience. The meeting between supervisor and employee to review the form often results in little or poor communication, each being eager to get it over with while avoiding confrontation over results or goals.

What is needed instead of an annual performance appraisal is a continuous effort in counseling and open, honest communications between supervisor and employee, coupled with continuous opportunities for relevant training. Group objectives (and a corporate mission) must be formulated and clearly communicated so that people in all departments and functions work as a team to meet customer needs. The communication needs to focus on the aspects of the employee's behavior that need to be modified. Regardless of the technical results achieved, it is typically the person's behavior that determines long-term effectiveness within the structure of a group. The one thing clearly under the control of the employee is his or her behavior. Relating behavior to how he or she performs their job is an essential and missing piece of typical performance appraisal systems. A prime evaluation criterion becomes promoting teamwork, not individual results. The focus becomes continuous improvement of the process, not single-mindedly pursuing individual numeric objectives. It is very important that the differences between common and special causes of variation are understood before you begin a system of more continuous feedback to employees. If feedback on common causes outside of their control are given to employees, you'll generate more drastic variations in performance. Employees will react to each piece of feedback, so it is critical that feedback relates to the special causes that they control.

A technique that has been successfully used as part of a continuous feedback system is called the "critical incident" method. It is based on the fact that every employee needs to know about behavior that is good, as well as that which needs modification. Such incidents occur randomly and are judged by the manager involved as they occur. When the manager is aware of an incident in either category, the incident is discussed with the employee. The manager writes notes of their discussion, gives a copy to the employee, and places the notes in the employee's file. With this technique, there are no surprises because the notes record information already given to the employee. Both the manager and the employee know what aspects of behavior are being appraised. It avoids talking with the employee based on perhaps less-than-accurate memories of past incidents. In the past, such notes were prepared only when someone's behavior or performance is so extreme that termination is being considered.

Adopting the approach as part of a routine system for continuous and effective coaching is usually greeted with some resistance, based on fears of continuous reporting and concerns about "keeping book" on people. When the company requires such continuous documentation as the basis for salary adjustments, the process is accepted, even if reluctantly, and the reality of the process soon dispels these fears. The process does compel managers to address employees on a face-to-face basis when an incident (good or bad) occurs, a responsibility that most avoid under the annual performance process.

Get away from job descriptions for employees and instead define the desired end result. If you cannot describe some tangible output, the job should not exist.

James Giardina

Another major barrier to quality work is a system of rigid job descriptions or classifications of people within a company (i.e., similar to the U.S. government system of GS grades with multiple divisions or steps within each grade). In companies with classification systems, you could be classified as an Engineer II, grade 6, for example. Typically, the classifications are tied to salary ranges and are based to some degree on education, years of experience and the types of duties carried out. Since years of experience are easier to define, you soon find a lot of weight given to seniority and experience rather than more subjective aspects, such as responsibilities met or performance. Professional societies often report their salary surveys with a similar scale. Some people are very comfortable without a structured classification system, while others are not. The ones that are not tend to be the average performers. The structured system, as does the annual appraisal system, fosters mediocrity. Try telling the super-performer that his or her raise is limited because his or her salary is at the upper limit for Engineer II, and you can't move from Engineer II to Engineer III until you have two more years experience. You'll soon be watching him or her pack up his or her books and leave. If you give him or her the higher reclassification and salary without the years of experience, you'll find others stuck in the Engineer II classification wanting an explanation. In a professional service organization, classification systems will decrease the overall quality of work. Coworkers are much more willing to accept differences in salaries without a classification system because they then base their judgments on their perceptions of performance, not on an artificially generated classification system label. Age and experience become less important than results when there is no rigid classification system. Pay for results promotes the flexibility that teams need to respond to changing conditions.

Focusing on each line item in a project budget can be equally harmful to project quality. You need to be flexible enough to move resources from one part of the project to another to manage resources as a project evolves, in order to optimize the end result. We've worked as consultants to large municipal governments that establish project budgets broken down into many task budgets. As the project proceeds, it may be clear that the initial allocation to the tasks was not proper. The paperwork required to change a task budget, even though it doesn't affect the total budget, is so burdensome that consultants are reluctant to even propose a change, although it would clearly benefit the project. Allow flexibility for the project team to shift resources to where they are needed to produce quality results. Don't let the financial management structure of your business or a given project be a barrier to quality work. Managing by the bottom line can also cause you to make decisions on data that may not be directly related to serving your customers' needs. It can lead to preoccupation with short-term results that don't serve you or your customers' long-term needs.

Close contact between management and the work force is essential in creating the teamwork needed for quality. On a peak performing team, everyone knows what is going on. Information must be had by all so that constructive feedback can occur and appropriate adjustments be made. When barriers are removed, the project manager moves away from solving problems on a one-to-one basis toward facilitating collective solutions involving everyone on the team that has a contribution to make. You become more of a coach and facilitator and less of a number cruncher. You concentrate on creating an environment where the team has the resources and systems needed to function efficiently.

Systems Concepts

An important concept in TQM is the differentiation between results caused by individual variations and those caused by features of the system. These are also referred to as special and common variation, respectively. Individual workers can do nothing to change the variation caused by the system itself—only management can. It is crucial that management realize this and not hold individuals responsible for such variations. It's a sure way to demoralize the work force.

Here's a simple example to illustrate the concept. Let's assume that your project team's assignment is to roll dice, and the goal is to roll as many 7s as possible. After the first hour, the following results were obtained:

Person	Number of 7's Per 100 Rolls
Nancy	32
Mary	27
Bill	25
Tom	23
Ed	17

An unenlightened project manager could look at these results and conclude that Ed is a marginal performer. Either he doesn't know what he's doing, doesn't care, or maybe has a drinking or drug problem. He's probably in need of counseling or training if he's even worth keeping around. Maybe Ed's not using the right procedures—better check out his methods and approach. Nancy is a sterling performer—probably should give her a raise, obviously a very talented person that works hard to produce superior results, should get more of the staff to emulate her. Mary, Bill, and Tom are average performers, cost of living raises only for them. The top two performers are women; maybe men aren't as well suited for this kind of work.

Of course, in this simple example, it is clear that the natural and random variation of the "system" (rolling of the dice in this case) accounts for the variation in individual performance. The individuals have nothing to do with the variations in the result. Reprimanding Ed will be counter-productive. It will demoralize Ed and produce no improvement because Ed has no control over the system variables affecting the results. Analyzing what Nancy is doing will also be a waste of time for the same reason.

In another example, you might identify all of the company's projects that were over budget. One approach is to hold each of the project managers personally responsible. You could let them know that the day of reckoning would come when the next salary adjustments were made if they didn't straighten up their act. Another approach would be to carefully look at the project losses. Looking at the list of projects, you might find that the dollar magnitude of only 2 of 25 project losses fell outside a narrow range of dollar overruns. In other words, there seemed to be two exceptionally large overruns that may well have been caused by special circumstances, but the rest seemed to be all of similar magnitude. It may be found that the similar losses were occurring in departments that were chronically understaffed—work was simply coming in faster than staff could be hired. As a result, people were being pulled off projects to fight the fire of the day, usually an imminent deadline. There was lost time involved in dropping one project, getting up to speed on another, and then restarting on the first project. By looking at the system, chronic understaffing would have been seen as the underlying cause. In the first approach, the company president failed to look at what in the system was contributing to the project losses. This "taking names and kicking ass" approach is inappropriate.

With these examples, you can begin to see the differences between the systems perspective and one that focuses on discrete events. The types of questions in each approach look like this:

Discrete Event Focus	Systems Focus
Who is responsible?	What in our system allowed this to happen?
Why did it happen?	How can we change the system to reduce the chances of recurrence?

The final outcomes on any project work are, of course, a combination of discrete contributors (such as an individual's effort and skill) plus the effects of the system. Dr. Deming's experience is that *at least 85 percent of the opportunity for improvement comes in the area of system improvements*. It is our experience that most managers look at the system's effects last, if at all. It is often easier to blame individuals rather than fix the system problems. Such an approach misses the target and needlessly destroys the personal relationships needed for quality work.

Rising from the discrete and systems effects are two types of variation: "common cause" from system effects, and "special cause" from discrete events. Their characteristics are:

Special Cause	Common Cause
Uncontrolled	Controlled
Does not belong to the system	Belongs to the system
Aberrant	Inherent
Unpredictable	Predictable

Often, it is possible to find the source of a special cause and alter or remove it so that it no longer affects the system. In some cases, this isn't possible. For example, Florida Power and Light found that one special cause—lightning strikes—was causing a substantial number of power outages. There was nothing they could do to eliminate the cause. In this situation, work on the system so that it is less sensitive to the special cause. They installed more lightning rods and added system fuses to prevent the effects of a lightning strike from traveling through their system. By altering the system, they reduced lightning-related outages by a factor of four.

Understanding the difference between common and special variation is essential to effective management of quality. Examples of common causes that management could correct include inadequate training, inadequate number of staff, inappropriate staff composition, overcrowded office space, lack of proper equipment, or staff scattered at several locations. Management must understand that unless they change the system, the system's capability will not change.

The basic steps then in approaching problems from a systems viewpoint are:

• Identify the problems.
• Observe the features of the problem.
• Find what in the system are the main causes of the problem.
• Eliminate the causes.
• Confirm the effectiveness of the change.
• Standardize the new procedure—change the system!
• Review the effects and plan for the future.

Unfortunately, these logical and effective steps are all to often replaced with blaming individuals and using wishful thinking, rather than collecting the data for proper analysis.

The message of the systems concept: managers should focus on gathering data and analyzing the features of the problem, to improve the process that produces results, rather than making judgments of the results. Management by human impulse responds to problems by looking for who screwed up. Management by the systems approach collects and analyzes data to see how the system allowed the problem to happen.

Statistical Process Control

Statistics are like a bikini. What they reveal is suggestive, but what they conceal is vital.

Aaron Levenstein

Statistics, while important, are just one of the tools in TQM concepts. A carpenter knows how to efficiently use a saw, but without a plan he can't efficiently build a house. Similarly, without a plan, statistics aren't of much help in assuring quality. In TQM approaches, statistical techniques are used to:

- Assist in determining why variability is occurring (i.e., why did it take 120 hours per sheet of drawings on project A versus 75 hours on project B?).
- Identify the causes.
- Change the process to reduce the variability.
- Monitor the results of the process change. Statistical tools, such as frequency distributions, histograms, and control charts, can be used.

The only (statistic) I can ever remember is that if all the people who go to sleep in church were laid end to end, they would be a lot more comfortable.

Mrs. Robert A. Taft

Once developed, statistical tools can be used by staff whether or not they have training in statistics. Control charts can be used to plot trends in the data to determine when something is about to go wrong on a process. If the first few sheets of drawings on a major project show a trend in hours used that is clearly above that experienced in the past, trouble could lie ahead. The control chart tells you it is time to take action before the project budget is in deep trouble. Control charts are also useful in tracking long-term system problems. If, for example, the number of hours per sheet of drawings shows a steadily increasing trend from project to project, it is likely there is a system problem that needs correction. Statistics are a useful way of tracking and demonstrating the value of the TQM efforts.

Suppose that data collected on the effort required to produce drawings for a construction project showed the following average distribution of labor tabulated at the top of page 223.

The amount of total time per sheet and the time spent in revisions appears unusually high compared to company experience. In checking the procedures, we found that, in this case, the project manager assigned the design work and then did

	Hours/Sheet of Drawings
Structural Engineer	6
Civil Engineer	6
Mechanical Engineer	6
Electrical Engineer	6
Initial Drafting	12
Revisions	14
Total	50

no review of the work until the initial drafting was completed. By failing to review the work while it progressed, an inordinate amount of rework resulted. In this case, the procedure could be changed so that the project manager reviews the design work before it goes to drafting. Data would then be collected to see if the change produced the desired result—less total hours and less time in revisions.

Let's work through an example to illustrate how statistics can be used for process control. You are managing nine drafters, and you are eager to determine how the relative quality of their work compares. After all, it is time for evaluations and salary review. Who gets a raise? Who gets penalized? To do so, you decide to compare the number of mistakes that they have made per ten sheets of drawings over the past year (see page 242 for a discussion of operational definitions, such as "drafting errors") as one of your evaluation criteria. Each of the drafters have essentially the same responsibilities and all had approximately the same large number of opportunities to make mistakes. Data collected over the year show the following results:

Employee	Number of Drafting Errors/10 Sheets
1	10
2	15
3	11
4	4
5	17
6	24
7	11
8	12
9	10

A typical approach would be to reward the employee who made the fewest mistakes and penalize those who made the most. This approach is based on the assumption that an employee's performance is significantly different from another's and is a direct result of differences in individual abilities. So, you'd rank the employee's performance as shown in the table at the top of page 224.

A key consideration is whether or not there is truly a *significant* difference between the number of mistakes. Consider, for example, that in a group of nine,

Rank	Employee	Number of Errors
1	4	4
2	1	10
3	9	10
4	3	11
5	7	11
6	8	12
7	2	15
8	5	17
9	6	24

one of the nine will be in the top 11 percent and one of the nine will be in the bottom 11 percent *no matter what the difference* in their performance is. The fact that one of the nine has to be the worst performer does not necessarily mean that the worst performance is significantly different than the best. Remember the example of the dice rolling team (see page 219). There were marked differences in performance, but none were significant because it was the random nature of the dice rolling process that determined the difference. The dice rolling process is in itself a system that obviously controls the results. It is not intuitively obvious whether or not it is the system or the individual drafters' abilities that are controlling the results of your drafting department. A control chart can be developed to determine the answer. Such a chart can be used anytime when:

- A count is made of the number of occurrences of an event that had many opportunities to occur.
- The event was unlikely to occur at any given opportunity.
- The area of opportunity is constant.
- The occurrences are independent of one another.

Your count of drafting errors meets all of these criteria so that you can proceed with construction of the chart. The first step is to calculate the arithmetic mean, which is, in this case, 12.67. The next step is to calculate the standard deviation, which is 5.57 in this case. For a normal distribution, we know that 95.45 percent of all values will fall between the mean ± 2 standard deviations (SD) and that 99.73 percent will fall between the mean ± 3 SD. These values can be used to calculate "control limits." Many TQM practitioners regularly use ± 3 SD to define control limits; however, there may be cases where you will want to examine other control limits. The degree of control is established by the range you select. The chances are certainly excellent that any values that fall outside of ± 3 SD are worth examining, to see if they are resulting from special causes, because they constitute only 0.27 percent of all values within a normally distributed set of data. However, you may decide to evaluate the causes of a larger portion of the data. The use of 3 SD is well-suited to manufacturing applications because the data are generally fast moving. The widgets keep rolling off the assembly line and you can quickly accumulate data. In service industries, data on items such as the example's drafting errors accumulate more slowly. There is a risk that special causes could be overlooked if a 3 SD limit is used and only 0.3 percent of the data are examined as potential special causes. If you use 2 SD, you'd consider 4.55 percent of the data as potential special causes. The risk from using 2 SD is one of overcontrol. Some of the data points outside of a 2 SD limit may be related to common causes, not special causes. You'll have to examine each system to determine appropriate control limits. In our drafting example, let's use, as an example, the mean ± 2 SD. The upper control limit would be 23.81, and the lower control limit would be 1.53. If all values fall between these two control limits, you are considered "in control." In a system in statistical control, the system, not individual differences, is causing the result. Management is then responsible for improvement because the system or process is going the best it can with the resources and environment available. If a point falls outside of the control limits, it means that someone is not operating in the same system as the others (i.e., is a special cause) and there is a significant difference in that person's performance. Figure 9-1(a) shows what the control chart would look like for your drafting group. The performance of employee number 6 falls outside the control limit. That employee's performance is significantly different and needs your attention.

What about the other eight drafters? Are they operating in a predictable system (i.e., a state of statistical control)? To determine this, eliminate the out-of-control point (employee number 6) and recalculate the average, standard deviation, and control limits, based on the remaining eight. This recalculation reflects reality only if you identify and remove the source of the special cause. Figure 9-1(b)is a control chart based on the remaining eight. It is clear from this chart that the remaining eight employees are all part of the same system—that is, their performance is being affected by common causes. Their performance will improve only if you as their manager improve their resources or environment. In this example, the number of drafting errors would not have been a valid comparative basis for the eight drafters in your review process. Other factors should form the basis of any differential ratings and salary adjustments. If one of the group had been outstanding (i.e., fewer errors than the lower control limit), the number of errors would have been a valid factor in a merit salary raise.

What about employee number 6? Did those 24 mistakes occur by chance, or are they likely to continue? To answer this question, we must look at his or her performance over time. The results could look like one of the control charts in Figure 9-2. If employee number 6 were in a state of statistical control, the control chart would look like Figure 9-2(a). A chart like this means that the employee will predictably, on the average, make 24 mistakes per year. A situation like this presents a problem because the worker is in statistical control (predictable) at an unsatisfactory level.

Dr. Deming's observations are that it is usually uneconomical to attempt to

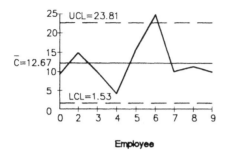

A. Control Chart, Drafting Errors for Nine Drafters

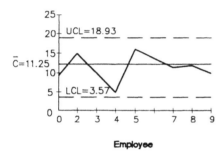

B. Control Chart, Drafting Errors for Eight Drafters
(Drafter Six Excluded)

FIGURE 9-1.   Example control chart—drafting errors by group of drafters.

retrain a person in this situation, but it is better to put the person into a new job in which training may produce better results. Further training in the same job is not likely to produce improvement.

On the other hand, if the data were not in statistical control (Figure 9-2(b)), the high number of mistakes might be temporary. The unusually high number of mistakes in September could be due to an illness, a unique emotional upset (i.e., death in the family), or some other special temporary cause. If this cause was indeed a one-time occurrence or can be found and corrected, employee number 6's performance could become satisfactory within the control limits. You should also look at the performance of the other drafters in September. If they consistently

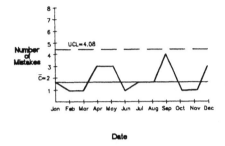

A. Drafter Six in Statistical Control

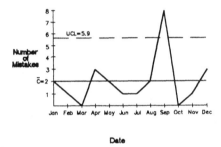

B. Drafter Six Not in Statistical Control

FIGURE 9-2.   Example control chart—drafting errors by employee 6.

show a higher-than-average number of errors in September, there was probably a system factor that affected all the drafters (e.g., unusually stressful work load, new type of drawings, introduction of new drafting standards). It may be a system factor that you should be alert for in the future. This example illustrates the importance of understanding variability in order to get consistency.

Statistics need not necessarily be limited to measuring totally objective parameters, like the number of drafting errors. They can also be used to measure changes in attitudes as the result of a change in management approaches. For example, the Paul Revere Company developed an approach to improve commitment to and enthusiasm for the corporate quality management program. It was called the Program for Ensuring that Everyone's Thanked (the "PEET" program). Each of the top 15 corporate executives were given the names of two quality team leaders to personally visit with each week. There was no set time to do it or topics to be discussed, but quality work was usually talked about. One vice-president said that he met people who had been with the company for ten years but that he could not remember meeting. There seemed to be a good correlation between improved attitudes about the quality process and enthusiasm for work in areas where the PEET program was most rigorously used.

Statistics could have been used to quantify the changes in attitudes as a result of the PEET program. The basic steps would have involved establishing baseline measurements before the PEET program, implementing the PEET program, and then making another set of measurements to compare to the baseline. Two types of variables can be used in such a situation: counting variables and rating variables. Counting variables in this case could have been the number of ideas per quality team per month and the dollar value of their ideas. These counting variables would indicate whether or not there were some quantitative improvements in quality team outputs. Attitudinal changes could also have been measured, using a rating variable. A rating variable could have been established, using a scale to measure the attitude of quality team members, such as:

1—The quality process is a total waste of time. There is no corporate commitment to quality.
to
9—The quality process is making an outstanding contribution to my job enjoyment and fulfillment. I know that the company management from the president on down is personally committed to this program.

A large number of quality team members could be asked to rate the quality program, using this scale. By statistical analysis of the counting and rating variables before and after the PEET program had been in effect for a few months, the program effectiveness could be evaluated.

Minimizing Variation from the Norm

Quality is not an act. It is a habit.

Aristotle

Higher quality means less variation from the norm. In terms of production methods, variation from the norm produces less uniformity of results and more defective work. A story often cited at TQM seminars illustrates this point. An American auto manufacturer used transmissions built to identical specifications from Japanese and American factories. The finished Japanese transmission had far better maintenance experience. Perplexed, since both were using the same specifications, the manufacturer tore down a dozen of the transmissions from the Japanese factory and a dozen from the American factory. They found that the parts in the American-made transmissions were within the tolerances from the specified value, but the Japanese-made parts deviated less from the specified value. For example, 99 percent of the American-made parts may have deviated, within 0.04″ from the specified value, while 99 percent of the Japanese-made parts fell within 0.02″ of the specified value. The deviations from the multitude of transmission parts act in an additive manner to result in the Japanese-made units operating longer and with fewer maintenance problems.

Consistent performance in a service industry is also important to long-term success. For example, consistent grouping of construction bids near the engineer's estimate does a great deal for the engineer's reputation. Bad news (like bids that are 30 percent over the estimate) travels much farther and faster than good news. It can take a long time to overcome the bad news from one underestimation on a major project. Wide variations of bids below and above the estimates are indicative of poorer quality work. In this example, statistics comparing actual bids versus estimated costs provide a basis to track performance and measure the effects of changing design or estimating procedures. The less the variation, the better the quality of the work.

Reducing variation leads to reduced rework, decreased cost, increased productivity, increased quality, and increased profit. As noted earlier, there are two causes of variation: common cause and special cause. Special causes result in the largest variation from the norm and require individual attention and correction. Overall process improvement depends on reducing common variation. For example, assigning an inexperienced or incompetent person to estimate the cost of a complex project or not allowing adequate time for an estimate would be two special causes that could result in large bid variations from the estimate. The cause is unique to that case. An example of a common cause of variation could be the use of a corporate unit cost data base that is updated quarterly, when a monthly update would track a changing economy more accurately.

Training

If you think education is expensive, try ignorance.

Derek Bok

A key to quality work is a corporate commitment to training on a broad scale related to technical skills, interpersonal skills, and quality management skills. Training in technical skills, law, and accounting will help technical people understand their jobs better. Training in communication skills, conflict resolution, psychology, consensus decision making, team building, negotiation, stress management, and time management will improve their interpersonal skills. With the person-centered and TQM approaches, interpersonal skills are crucial. Team members must communicate more effectively than in a structured environment. They no longer rely on the boss to coordinate communications and handle interpersonal conflicts. Day-to-day interactions can become chaotic and stressful unless training in the basic communication skills described in Chapter 2 is provided. Interpersonal skills become essential. Training managers to be more people-oriented pays big dividends. After all, project success is achieved through people, and people-oriented managers have the most success. Training on the theories and practice of TQM is a key step in a TQM program that everyone should receive (yes, *everyone*), from the very top to the very bottom of the organization. Managers need to continually ask themselves what their employees need to learn in order to do a better job.

I can't tell you how many times I've heard that expression, 'Our most important asset is our people.' But captains of U.S. industry air-condition their computers before they air-condition their people.

Richard LeFauve

"Employees are our organization's most important asset." This is stated so often that it is a cliche. One way to demonstrate that there is substance to this claim is to commit to a long-term program of training. When times get tough, cutting the training program sends the wrong signal, as well as being ill-advised, since training will produce improved performance at a time when it is critically needed. Garfield notes that peak performing companies often increase their training budgets when times are tough. Marginal companies will be cutting theirs because they consider the program as a fringe benefit. The peak performing companies focus on what they need in order not just to survive, but to excel, which is the surest way to survive. It is interesting to note that, according to Scherkenbach, Japanese managers are given flexibility to cut costs in many areas, but one area they cannot reduce is the training budget because training and education are the keys to consistency and performance. A proper training program provides improved morale, pride, enthusiasm, and an improved organizational climate that promotes better working relationships.

The effectiveness of training is limited by restrictions on using the training on the job. How many times do you see someone return from a training session to be greeted by a manager saying that a lot of work piled up and it's time to get back to work? Or, new employees are immediately thrust into the fire of production? One of the approaches used at the Paul Revere Company involved a half-day class on their quality process for every new employee. There were a lot of raised eyebrows and skeptical looks when they told the new employees that they welcomed their ideas on how to improve their operation. The concept is hardly standard (at least not yet) in the business world. Once the new employees realized that they truly did have power over their job, the approach was enthusiastically accepted. The training returned quick dividends.

A common barrier is the attitude that the training is for my staff, but not for me. Managers may feel that they do not need to know the details offered in a training, that they don't have time, or that they know it all already. Unfortunately, many managers are not knowledgeable about the system they are responsible for. Participation of the managers underscores the importance of the training, as well as giving them the same frame of reference as the work force—and possibly some improvement in their own skills! Other barriers result from attitudes of managers that it is too costly to train all staff, especially the support staff. Some managers have the attitude that training should be aimed at only the difficult people, to "fix" them. These attitudes are counter productive. Training applies to everyone in the organization.

Training should be considered as essential to the growth of individuals and the company itself. The training should be related to the jobs or personal needs of the employees. Presenting theory without the practical applications is a waste of time. The most successful training includes opportunities to apply the new skills on the job *immediately* after the training.

The more successful service firms have come to appreciate the value of training. Several are committing about 2 percent of their total fees to a corporate training program. IBM makes an educational investment of 5 percent of its payroll. Two Malcolm Baldrige award winners, Xerox and Motorola, have spent $2,500 and $1,250 per employee per year on training. Training can be an extremely effective vehicle for pulling the various elements of a company into a smoothly functioning team. Training becomes a permanent process, not an event.

Leadership

Be fanatics. When it comes to being and doing and dreaming the best, be maniacs.

A.M. Rosenthal

Much of Chapter 3 is devoted to the subject of leadership, and later in this chapter we will discuss the importance of management's leadership in achieving success in a quality program. Suffice it to say here that the universal experience in successful quality programs is that the commitment to quality emanates from the leader of the organization downward. Leaders truly must walk like they talk.

ANALYTICAL TOOLS

There are a few basic tools commonly used in the TQM approach that are useful in defining improvements to a specific process or resolution of a specific problem. The tools are applied in the following sequence:

1. **Flow charts**—To define the process being evaluated.
2. **Brainstorming**—To develop all possible causes.
3. **Cause and effect diagrams**—To categorize the possible causes.
4. **Pareto charts**—To determine the most significant of the causes.
5. **Determining root cause**—To ascertain what factors underlie the apparent causes.
6. **Brainstorming**—To develop solutions to the most significant problems and root causes.
7. **Cause and effect diagrams**—To categorize possible solutions.
8. **Pareto charts**—To determine the most significant solutions.

Flow Charts

A flow chart of a process is very useful in identifying all the steps involved and in visualizing the process. No matter how simple the process might seem, the people involved may have different concepts of the steps involved. It is interesting to have several people who are involved in a process independently develop flow charts. More than likely, you'll get back a wide variety of ideas of the steps in the process and their interrelationship. A flow chart gives you an overview of the steps involved, their sequence, what decisions are made, and how they affect the flow.

As an example, let's take the reasonably simple task of producing a written report for a client. We'll start at the point that the project work has been done and the task is to convert the work into a final, printed document for the client. Figure 9-3 presents how you might prepare a flow chart for the process. The diamond-shaped boxes are decision points, while the rectangular boxes are steps that add value to the process. The chart can then be used to facilitate discussion about what might be done to improve the process. It might go something like:

> Our reports have been going out with some typographical errors, even in the cover letters. We need to spend more effort on proofreading. Who is doing it now? Who should be doing it? Are we allowing enough time in the process for proofreading? After all, a sloppily prepared report gives the client a poor impression of our quality. Some of our reports are hard to understand. Should we add a technical editing step before it is proofread? We've had our clients point out that our reports haven't addressed all of the scope requirements. That's embarrassing. We need someone other than the author to check the draft against our scope of work. We haven't been allowing enough time in our schedules for client review, and we need to adjust future project schedules to do so. We've had some reports go out with pages missing or inserted upside down, so we need to make some changes in the final production step. We need to be sure to allow enough time to have someone (and let's decide who) go through each copy to be sure it is assembled properly. We've had some last-minute confusion on who is to receive copies. We need to send the client a draft distribution list along with the draft report.

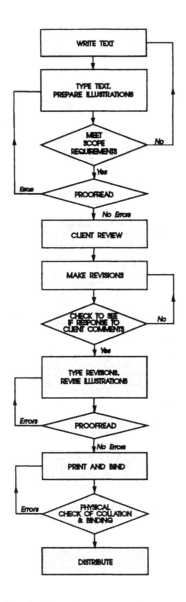

FIGURE 9-3. Sample flow chart for written report to client.

Here are some keys to effectively flow-charting a process:

- Draw the flow chart of the process as it exists, *not* as you would like it to be.
- Be sure to involve people who know what is really going on in the process. A flow chart must reflect reality, to be of the most value.

- Do not omit key decision points. A common error is to overlook all of the points in the process where decisions are made.
- Expect to redo the flow chart several times until it is complete and accurate. Have it reviewed by everyone with a stake in the process. A useful technique for getting reviews is to post the flow chart in a commonly-used area (coffee break room, for example), with a supply of blank self-stick notes. People are invited to write their comments on the notes and stick them to the flow chart. Even those who are normally reluctant to speak up will often provide comments with this approach.

Brainstorming

The process of brainstorming involves generating a large number of ideas on a specific problem, solution, opportunity, or process from a group of people. A key is to get the right representation on the team. The team should be a vertical slice through the project or organization involved—that is, if the problem involves production of plans and specifications, the team should include professionals, technicians, drafters, clerical, the office manager, and project managers. A homogenous, horizontal slice of just, for example, professionals will preordain a narrow viewpoint on the problem. Quality improvements come from involving everyone in the process, as discussed earlier in this chapter.

Brainstorming occurs in a free-thinking, safe environment in which ideas are generated and recorded *without any judgments or censorship*. A "wild" idea by one person can stimulate an idea critical to solving the problem from another. Ridicule or harsh judgement will ruin the creativity of the session. Generally, one person acts as the session facilitator and one as the recorder (see the earlier chapter on effective meetings). The facilitator has the job of keeping a freewheeling, open, synergistic environment and, often a tough job, of keeping all judgments and discussions of ideas stopped until the brainstorming process is complete. The recorder's job is to quickly and accurately write each idea down as it is generated. Using large flip charts allows the ideas to be seen by all, and the sheets serve as a group memory. The session should continue until no new ideas are presented.

To ensure equal participation, allow silent time for each team member to write down their ideas. An assertive team member may otherwise dominate the session and intimidate others. Then, ask one person for one idea. Each person is asked for an idea with no comments or judgments by others. Make it clear that it is alright to pass on one turn and then speak up on a later turn if an idea has occurred. Often, one idea from another person will trigger ideas from others. Repeat the cycle until each person passes.

To illustrate the concept, Figure 9-4 is a list of ideas generated by a team working on the problem of rather limited sharing of work between offices of a firm, even when some offices were overloaded and others weren't. This list will later be

---

PROBLEM: Ineffective Work Sharing Between Offices

Local office wants to build up staff.
Project staffing requirements are inadequately defined.
Local office need/drive for high utilization/profit.
Mistrust between offices.
Negotiations fail between departments on sharing revenues from job.
Lack of marketing forecasting/adequate measurement of staff forecasting.
People resources are not known.
Staffing at project location is poorly evaluated.
Client demand for local staffing.
Past poor experience with quality and/or cost.
Centralized knowledge of who's available, who needs work.
Inability of project manager and department manager to reach agreement on scope of work with outside office.
Work product not defined early enough to commit resources.
Capabilities of staff are not on common basis for departments, regions, nationally.
Lack of established one-on-one working relationships between staff at different locations.
Available resources not previously used when they were available, causes ill feelings.
No clear-cut policy on decision making for staff utilization or enforcement.
Project manager too demanding on capabilities.
Lack of readily-available, up-to-date staff data base.
Lack of communications.
Poor communications between project manager and department managers.
Past bad experiences between key players (personal).
Lack of corporate standards and guidelines for production.

FIGURE 9-4. Sample brainstorming output.

analyzed to determine which are the most significant ideas. Brainstorming can be a fun, energizing exercise for the team, as well as a very productive source of solutions.

Cause and Effect Diagrams

Brainstorming can generate a lot of ideas—even an overwhelming amount of information. Cause and effect diagrams, also referred to as fishbone diagrams, are often used to categorize and organize the information. The method of constructing a diagram (see Figure 9-5 for a sample based on the brainstorming shown in Figure 9-4) is:

- Determine the effect. This is the problem the team is addressing.
- Put the effect in a box on the right-hand side of the paper.
- Draw an arrow from left to right, pointing to the effect.
- Using the ideas generated in the brainstorming sessions, group the ideas into categories of causes. In the case of our example, the causes were people, environment, planning, policy/procedures, and tools.

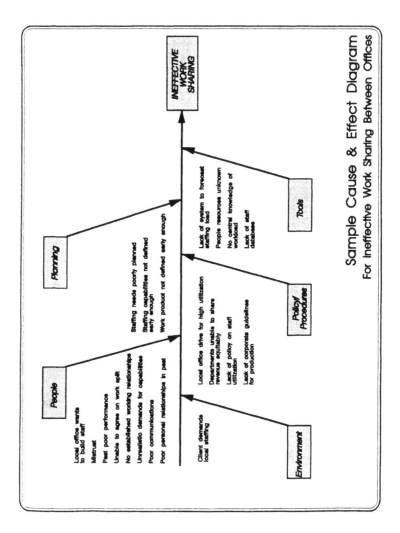

FIGURE 9-5.   Sample cause and effect diagram for ineffective work sharing between offices.

- Put the categories into boxes above and below the main arrow, and direct a branch arrow to the main arrow.
- Put the ideas into their proper category.

Once the ideas are grouped by category, the list of possible causes isn't as overwhelming. You may find similar ideas from brainstorming that should be combined into this one. Each category can be studied to find the most probable causes. The cause and effect diagram promotes group discussions of problems and displays information. It provides a summary with much greater clarity and eye appeal than the flip charts of ideas produced by brainstorming. There is also synergy in this process, resulting in additional ideas when the group focuses attention on the diagram. By categorizing the ideas, it becomes easier to do the next step—determine which ideas are most significant.

Pareto Analysis

Vilfredo Pareto, a nineteenth-century Italian economist, noted that the most significant items in a group normally are a relatively small portion of the total items in the group. The purpose of this step is to separate the important few elements from the trivial many by constructing Pareto charts. These charts simply arrange the data in a manner that clarifies the important elements and may suggest something that otherwise might go unnoticed. The chart focuses everyone's attention on the few aspects that may pay the largest returns.

Construction of a Pareto chart is relatively straight forward, especially if the issues being addressed can be defined by real measurements. To construct the chart, use the following steps:

1. Identify the feature or issue you are studying, such as the cause of change orders in this example.
2. List the categories that contribute to that feature. For this example, we list the various causes, such as owner changed the requirements, site conditions changed, specifications or plans were in error, and so on.
3. Tabulate the frequency of events for each category. List the categories in descending order of frequency.
4. Construct the Pareto chart with the frequencies plotted in descending order.

For example, let's assume that the issue under consideration is the cause of change orders on projects. Several completed projects have been analyzed to determine the cause of the change orders with the following results shown in the table on page 238. Figure 9-6 on page 239 is a sample Pareto chart for the change order causes.

The Pareto method can also be very effectively used to determine the most likely causes of a problem for which there is no hard data. Let's return to our

Cause	Number of Change Orders
Owner changed requirements	28
Changed site conditions	16
Specifications or plans in error	14
Equipment substitution proposed by contractor	8
Weather impacts on schedule	4
Manufacturer unable to deliver specified equipment	3
Delays due to processing change order requests, requests for information, shop drawings	2

example on work sharing between offices. The team that brainstormed and categorized the causes (Figures 9-4 and 9-5) was made up of a vertical cross section through the company's organization chart—company president, national corporate headquarters staff, regional managers, local office managers, project managers, professional project staff, and drafters. As discussed earlier, this heterogeneous team composition is important in order to get several perspectives on the issue.

The likely, relative significance of the various causes identified by brainstorming can be quantified by using the "Ten-Four" method. Each person on the team is given a total of ten votes and he or she must cast all ten of the votes, distributed among the identified causes according to that person's perception of the relative importance of each cause. No more than four votes can be given to any one item. If a team member feels there are only three important causes, he or she might cast four votes for the one he or she feels are most important and three votes for each of the other two. If the person feels that there are several important causes, the votes may be distributed 3-2-2-1-1-1, for example. The team facilitator calls out each item on the cause and effect diagram. Each team member votes on the item by holding up one to four fingers. The facilitator sums the votes for the item and records the sum on the cause and effect diagram next to the item. This technique develops a group opinion on the few important causes, as opposed to the many trivial causes. The results of the "Ten-Four" voting for the work sharing example is shown in Figure 9-7 on page 240. Of the 23 items on the original brainstorming chart, 5 have now been identified, which represent about 60 percent of the team members votes as the most important causes. However, at this point there may only be the apparent causes of the problem, as opposed to the root causes.

Determining Root Causes

Once the most apparent causes have been identified, further digging may uncover the root cause underlying the apparent cause. A technique used for this purpose consists of asking "why" about the apparent cause at least five times. For example, let's use the "poor communication" cause from Figure 9-7 (see the table on page 240).

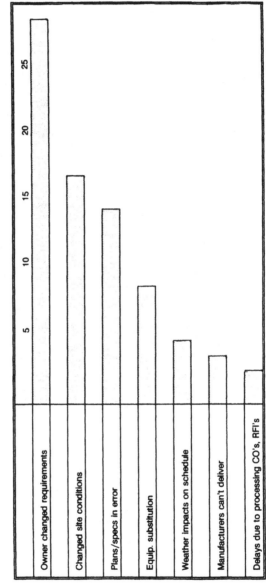

FIGURE 9-6.   Sample Pareto chart.

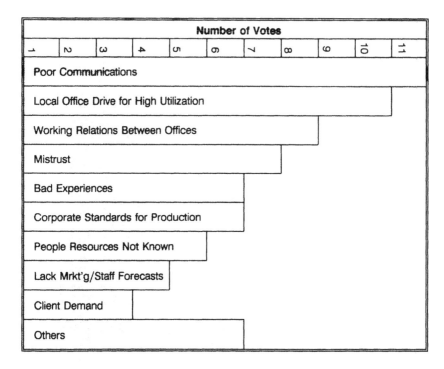

FIGURE 9-7. Sample Pareto diagram—ineffective work sharing between offices.

There may be other avenues to explore on poor communication that may lead to different root causes. There is no right or wrong answer—all may provide valuable information. It is normally easier to find solutions to the root cause rather than the apparent cause because it tends to be a more specific and manageable task to tackle.

Sensenbrenner offers an example of getting to the root cause. After becoming mayor of Madison, Wisconsin, he instituted one of the first TQM programs in public government. One of the problems that caught his attention was the difficulty in keeping city vehicles maintained. He started by talking to those closest to the problem—the mechanics at the city garage. When asked *why*, they said there were simply too many different types, makes, and models of vehicles (there were 440!).

Poor Communication (Apparent Cause)

Why: Department manager doesn't advise other departments of needs.
Why: Not sensitive to the benefits of work sharing.
Why: Corporate position and policy on work sharing not clear.
Why: Corporate position and policy not disseminated or explained.
Why: Corporate position and policy doesn't exist (root cause).

When asked *why*, they said that their parts purchaser said it was city policy to buy whatever vehicle had the lowest sticker price on the day of the purchase. Considering all the downtime and poor resale values, the policy didn't make sense to the mechanics. The mayor went to the parts purchaser to ask *why*. He agreed it would be better to have fewer parts to stock from reliable suppliers. But, he said, central purchasing wouldn't allow it. On to central purchasing with the next *why*. Central purchasing said that the comptroller wouldn't change the policy. When asked *why*, the comptroller agreed that the logic for a change was strong, but that the city attorney would not let him do it. Obviously, the next *why* was directed to the city attorney. His response was that, of course, you could make a change if the specifications included the cost of maintenance, availability of parts, and resale value. The attorney said he assumed the city was doing that all along! By asking *why* six times, the root cause and solution were found.

DEFINING AND SELECTING MEASUREMENTS

To determine the effectiveness of changes implemented in a system, it is essential that you measure the starting point and any changes. You can't apply statistical techniques until you have measurement data!

Designing Experiments

One of the first steps after identifying the issues is to define the basic question: "Will the process (or product) improve as a result of my formulating and carrying out a particular strategy?" You will need to choose certain factors for study, vary these factors in a controlled fashion, observe the effects of your actions, and make a decision based upon the results.

Choosing the factors to be studied is an important step. Often, experiments are based on two or three factors selected by technical personnel. The risk is that they may overlook other important factors or the interrelationships between the selected factors and other variables. Again, use the vertically-integrated team approach described earlier. Brainstorming the potential variables with this type of team may result in a list of 20 to 30 items. Using the Pareto method, a consensus can be reached on the most important variables to be measured.

Key Measures

Important criteria for measuring the key variables are that they should: 1) be directly related to meeting a customer's needs (i.e., driven by the needs of the downstream customer); 2) be measures of quality; 3) be measurable in a quantitative manner; and 4) be part of a regular, systematic feedback system with data to tell you if you're meeting the customer's needs. Feedback is a vital part of the process. People will soon lose interest without it. The measures may well change as you learn more about the process and the customer's needs. You couldn't make adjustments (or maintain your interest) in your golf game if you weren't allowed to watch the flight of the ball after you hit it.

The variables to be measured can be either counting or rating variables, often both. A counting variable might be the number of drafting errors per sheet of drawings. A rating variable could be the severity of the error. Rating variables tend to be subjective, but can still provide much learning. If we find the rating variables to be moving in a positive direction over time, we can conclude that there has been a positive change. For errors on plans and specs, a rating scale could look like:

9—Very serious error, creating a hazard to people, a functional failure, or a design that can't be built.

7—Serious error that would probably cause a hazard to a person not familiar with the project; likely to cause a functional failure; certain to cause substandard performance, increased maintenance, or undue construction effort.

5—Moderately serious error that could possibly cause functional failure, construction problems, substandard performance, increased maintenance, or extra construction effort.

1—Not a serious error because it causes only minor deficiencies in appearance.

After defining and choosing the variables, a baseline must be established, prior to any changes. If historical data are available, they can be used for the baseline. For example, if the number of hours spent on each sheet of plans for initial preparation and rework have been logged over a significant period of time, a baseline can be established from existing data. If data related to our variables are not available, we need to measure them over a specified period of time, to establish a baseline before making changes. Changes can then be made, and another set of data collected to compare with the baseline to evaluate the effects of the changes. Ideally, data should go directly to those doing the work. They can then track their progress, find solutions to problems, and will feel less threatened than they would if their managers first processed the data.

Operational Definitions

It is important that the variables be defined. This seems obvious, doesn't it? However, a variable as apparently straightforward as "drafting error" needs to be defined. Two people may have different definitions and get a different count of errors on the same sheet. An example definition for drafting errors could be: omission of all or part of a detail, incorrect dimension, inconsistent cross-reference, disagreement between drawing dimension and product schedules, or sheet format and standard drafting guidance (e.g., line weights, lettering) violation.

Example Measurements

Although services may not appear as amenable to measurements as manufacturing is, there are several measurements that can and have been used in TQM programs. These include:

- Turnover rate of technical staff.
- Claims, settlement, and litigation expense.
- Success rate on proposals submitted.
- Cost of marketing as a percentage of total fees.
- Client rating of firm.
- Number of schedule milestones missed.
- Amount of overtime by staff.
- Number of times a document is changed after it is issued.
- Dollar amount of field changes.
- Cost of drawing/specification rework after final check.
- Number of inconsistencies between drawings and specifications.
- Hours and/or cost per drawing.
- Number of contractor requests for information.
- Change orders, expressed as percentage of project construction costs.
- Ratio of final construction cost to estimated construction cost.
- Ratio of project engineering/architecture cost to budget amount.
- Submittal (shop drawings) review time.
- Response time to contractor requests for information.

An example is offered by a TQM team evaluating the drafting procedures in a design firm. The team selected the following key measures and operational definitions:

Key Measure	Operational Definition
Accuracy	Errors per drawing
Productivity	Hours per drawing /Cost per drawing
Efficiency	Rework hours per drawing
	Number of rework cycles per drawing
Timeliness	Days plus or minus from scheduled completion

You must know what you are going to measure and how you are going to measure before you can collect the right data.

INCORPORATING TQM CONCEPTS INTO MANAGEMENT PRACTICES

Cost of Quality

First we will be best. Then we will be first.

Grant Tinker

One of the first and most obvious questions to be addressed is whether or not the cost of making changes to improve a quality program can be justified. Prior to implementing a TQM program, costs increase, while quality, productivity, and competitive position remain stagnant. After an appropriate TQM program is implemented, costs will begin to decrease, quality and productivity will begin to increase, and, eventually, competitive position improves. No universal time scale can be applied. Patience is needed because the results will not be instantaneous. In some cases, it is months or years before a substantial difference in competitive position is confirmed. This is a lesson that Japan's success offers. Their position in international markets soared when quality improvements were realized.

Service organizations spend about 40 percent of their operating costs doing things over. When you get people doing things right the first time, you save all that money.

Philip Crosby

The value of successful TQM is apparent if you consider the relative cost of doing a job right the first time versus making a mistake, finding the mistake, tracking the defective work, and fixing the mistake. Big project losses come from the costs of not doing the job right the first time. The client's perception of the quality of your work will be boosted dramatically when the work is done right the first time. If erroneous work reaches your client, the error costs not only immediate dollars, but jeopardizes demand for future work. This has led to what some call the "1-10-100 Rule": if the cost to prevent a defect is $1, the cost to catch the defect by inspection is $10, the cost if the defective work reaches the customer will be $100 because not only will you lose that customer's business, but they will tell others. The following quote further illustrates the point:

The earlier you detect and prevent a defect, the more you can save. If you throw away a defective 2-cent resistor before you use it, you lose 2 cents. If you don't find it until it has been soldered into a computer component, it may cost you $10 to repair the part. If you don't catch the component until it is in the computer user's hands, the repair will cost hundreds of dollars. Indeed, if a $5,000

computer has to be repaired in the field, the expense may exceed the
manufacturing cost.

Richard Anderson, General Manager
Hewlett Packard Computer Systems Division

One approach to assessing the value of TQM is to look at the cost of nonconformance. It is not quality that costs, but rather nonconformance that is expensive. The costs of quality become very reasonable when compared to the cost of not achieving quality. An excellent example is offered by consulting engineering and architectural firms. Based on our discussions with many such firms, it is very common (nearly universal) for the firm's new work to be obtained at prices that include a profit that is higher than what will be eventually realized by the firm when the project is completed.

Let's use a hypothetical example to illustrate the point. A 500-person firm obtains $40,000,000 in new contracts over the course of one year. The policy and historical practice of the firm has been that the pricing of the contracts includes an average 12 percent profit, or $4,800,000 in total. At the end of the year, the firm's profits are 6.5 percent, or $3,120,000. These percentage profits (12 percent included in contracts, 6.5 percent actual profits realized) are fairly representative of many firms in the U.S. today. In fact, some of the larger firms we've talked to would be pleased to reach the 6.5 percent profit level even though their pricing is based on 10- to 15-percent profit. In our example then, the cost of not achieving quality is $4,800,000 minus $3,120,000, or $1,680,000. This cost not only illustrates the cost of non-quality, but also provides a useful long-term measurement of the effectiveness of a TQM program as it is put into practice.

Achieving quality is inexpensive when compared to the cost of not achieving quality. You may, however, encounter resistance to the cost of a new quality program because the costs of beginning and continuing changes to improve quality are new, easy to quantify, and readily apparent. Unfortunately, the costs of not achieving quality are costs that people are familiar with and accept—cost of redoing work, project budget overruns, missed schedules, and so forth. Specific project losses over a period of time will be a useful parameter to track. As you consider implementing practices and policies to improve quality, it will be important (and not necessarily easy) for you to keep this perspective on the relative costs of quality versus non-quality. It will also be important that everyone in the organization share this perspective.

A story from World War II offers an unusually graphic illustration of getting everyone to buy into the concept of effective quality control. Those who packed parachutes were told that they would, on a random basis, be required to make jumps occasionally with the parachutes they had prepared. The potential cost of non-quality became very real and understandable. They had no problem with becoming involved in quality control or relating to the needs of their customers!

Management must have a *long-term* obsession with quality work and continued improvement if gains are to be made on the quality of work. Short-term costs to implement changes to improve quality must be looked at as steps in a long-term process that will return dividends many times greater than these costs. The focus should be on what improvements in the quality of the work will be achieved over the next several years, as opposed to next quarter's profits.

It takes less time to do a thing right than it does to explain why you did it wrong.

Henry Wadsworth Longfellow

IMPLEMENTING A QUALITY IMPROVEMENT PROGRAM

Implementing a Deming quality strategy is not simply a matter of adopting a new set of slogans or a new accounting system. It's a matter of radical restructuring—part sociology, part systems theory, and part statistics—all aimed at liberating human ingenuity and the potential pleasure in good work that lies at least partially dormant in every organization.

Joseph Sensenbrenner

The Essential Ingredient

Read my lips. No new taxes.

President George Bush

People watch your feet, not your lips.

Charles Harwood
President, Signetics Corporation

The first and essential step is that the top manager makes a deep commitment to quality improvement and becomes personally active in the pursuit of the goals of the program. A lot of preaching without the personal investment of a substantial effort in the program won't work. Dr. Deming began his campaign in Japan by insisting that the president or chief executive officer be involved in the quality improvement efforts. The CEOs began walking the production areas, interacting with production personnel, and being open to staff input. The results speak for themselves.

He does not preach what he practices 'til he has practiced what he preaches.

Confucius

Management must demonstrate that long-term improvement is its highest priority, or the improvement will not occur. Management can demonstrate this priority primarily by its own action, as well as by expecting improvement from everyone and every unit in the organization and by providing the necessary resources for improvement to occur. Management provides leadership by giving clear, consistent direction to the program; establishing a firm, expeditious decision-making process; providing the necessary instructions, equipment, and materials; performing regular assessments of the program; and assuring prompt and effective correction of problems.

As discussed earlier, a successful quality improvement program was instituted in a large service company, the Paul Revere Insurance Company. The program began with a memo to every employee from the chairman of the board of the parent company. The memo stated the commitment of the top manager to the essential elements of a successful program. The memo stated that it was a key corporate objective to be identified both in perception and in fact as a high-quality, high-productivity company in all its business activities. The credo expressed in the memo included:

- High quality and value of the services for customers are the basic building blocks of the corporate strategy and critical components of its mission.
- Maintenance of high-quality, value, and high-productivity standards are at least equal in importance to corporate profitability.
- High quality and sustained productivity improvement are vital to achieving and sustaining its competitive advantages.
- High quality and low costs are linked and fully compatible objectives.
- Competent employees are the most important company resource and can make the largest contribution to maintenance and improvement of high-quality and productivity goals.
- Sustained corporate-wide activity, vigilance, and concern with respect to high-quality and productivity goals are a permanent part of the corporate culture.
- High quality means defect free, in conformance to requirements, and doing it right the first time—nothing less expected.
- Productivity means the rate of total true useful output produced per actual unit of total resources expended—simply put, the ratio of output to input.
- Value means the assessment by the customer of their satisfaction with the degree of excellence and the fair return they receive in products or services for their payments to and relationships with the company.

Always remember the distinction between contribution and commitment. Take the matter of bacon and eggs. The chicken makes a contribution. The pig makes a commitment.

John Mack Carter

Experience has shown that nothing short of the personal commitment and involvement of the top manager to such concepts is needed for successful quality improvement. Failure to do so results in the perception that top management is not really concerned about quality. The staff will question why they should be interested in quality if management is not.

Management must believe in the basic concepts discussed earlier:

- The role of management is to help its people do a better job.
- Good performance originates from people's inherent desire to do a good job and take pride in their work.
- The best sources of ideas for improvement are the people working in the process.

This belief can be demonstrated by management by:

- Expecting and requiring continuing improvement by getting everyone involved in improving projects.
- Establishing a clear process for determining improvement priorities.
- Constantly looking for opportunities to solve problems through teamwork.
- Constantly promoting awareness of customer-supplier relationships.
- Providing coaching, counseling, and training.
- Asking the right questions: Who is the customer? What do they need? Are we improving? How do we know? Where's the flow chart?
- Providing the resources, such as time for training, time for planning, time for quality team meetings, investment in systems to collect measurement data, and recognition of those who contributed.

We are aware of one firm where the president scheduled all quality team meetings at the same time. He personally patrolled the hallways, to make sure his staff attended, and he spent some time in each meeting. This is a graphic illustration of the top manager walking the talk.

The Key Steps

Figure 9-8 presents a diagram for developing and implementing a quality improvement program. The key steps are discussed below.

Commitment of Management

As we say in the sewer, if you're not prepared to go all the way, don't put your boots on in the first place.

Ed Norton
"The Honeymooners" TV show

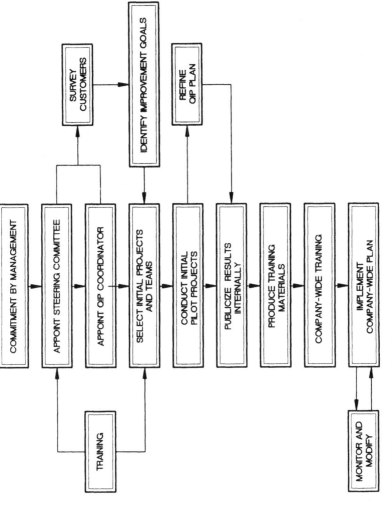

FIGURE 9-8. Development of Quality Improvement Process.

As discussed in the preceding section, this is the first step, and an essential one. Without it, there is no need to go further. In addition to the steps discussed above, management can demonstrate its understanding, priority, and commitment by training its management team first. One large firm, once committed to the program, sent 36 top management people to a four-day quality improvement training program. This sent a clear message on the commitment to the program.

Establish a Steering Committee

The steering committee should be chaired by the top manager. Its purpose is to provide direction and resources for the quality improvement program. Steering committee members should be inquisitive, open to change, creative, and able to interact effectively with all levels of the organization. It's generally best to start small and build the committee. Often, the committee starts with the top manager, the human resources director, and a few (two to four) top technical people that must support the system changes. The committee members need training on TQM concepts. They should visit others on-site in similar organizations that have experience with TQM concepts.

   The steering committee responsibilities include:

* Develop a statement of the organization's mission and guiding principles.
* Develop an action plan for implementing the program.
* Select a program coordinator.
* Emphasize how changes can take place to create an environment that encourages everyone to focus on continuing improvement.
* Identify improvement goals.
* Select two to three initial quality improvement projects and teams.
* Ensure that the teams get the required support.
* Appoint and provide training for liaisons between teams and steering committee.
* Review team recommendations, and integrate them into the quality system.
* Decide if any specialized support is needed.
* Promptly distribute news of success stories from inside and outside of the organization.
* Establish a system that gets accurate feedback on the program success.
* Be visible and available to all employees, and ask the right questions.
* Conduct an internal survey, to determine how people feel about their environment now and to get ideas for improvement.
* Conduct a survey of clients, to determine their impression of the services provided and areas that need improvement. Customer surveys will provide valuable information and often unexpected results. A division of Hewlett-Packard was having difficulty with receivables—12 percent of their receivables had been due for more than 90 days. They assigned a TQM team to the problem. The team's initial assumption was that a substantial part of the problem came

from customers with cash flow problems or who were simply avoiding paying their bills. When they surveyed their customers, they found that nearly all of the delays were due to problems caused by Hewlett-Packard: incorrect invoices, shipment problems, invoices sent to the wrong person or location, and so forth. By addressing their own invoicing procedures, they reduced the problem receivables from 12 percent to 1 percent. Be aware that when you survey your customers, you will create an expectation that things will be changing. If you don't follow through with an improvement, you will have harmed yourself with your customers.

Appoint a Coordinator

As noted above, the steering committee should select a coordinator who will:

- Monitor acceptance of the program and give feedback to the committee.
- Give advice to committee on resources needed.
- Work closely with the TQM teams and liaisons, to provide resources and do whatever is necessary to ensure that the teams succeed.
- Be the link with outside firms (i.e., keep in touch with peer organizations who are implementing or have implemented similar programs) and with major internal functional groups.
- Assist the committee by scheduling meetings, distributing minutes, updating the program plan, and bringing significant articles or other information to the attention of the committee.

Provide Training in Quality Improvement to TQM Team Leaders

To be successful, a quality improvement program is going to involve everyone in the company and a significant amount of their time in quality team meetings. For example, the successful Paul Revere Insurance Company quality program involves everyone in the company on a team. Each team meets for 30 minutes a week. The meetings have to be extremely efficient in order to be productive. Considering that there are about ten people per project team (larger teams can lose someone's ideas in the shuffle), many team leaders are needed. Very few of the leaders are likely to have experience at leading effective meetings or leading groups through problem solving. Thus, training becomes critical. An approach that was very effective for Paul Revere was to have a dozen or so of their employees trained by outside experts to be quality instructors. These instructors then taught the team leaders. The advantages of this approach over outside training include:

- Emphasizes that the quality process is an in-house effort. "Top-down commitment and bottom-up implementation" has been used to describe the successful corporate attitude.

- The instructors are able to relate quality concepts directly to the staff's experience.
- The training can be made part of the regular training program, and a consistent basis for training all team leaders can be maintained.

In the Paul Revere case, each team leader attended three full-day training sessions in a month, followed by one session a month for the next three months.

Selecting Initial Projects

Before proceeding with company-wide implementation, it is productive to select a pilot project, to establish the credibility of the program. There is nothing like an internal success story to generate interest! By "project," we do not necessarily mean a specific design or construction effort (like the design of a bridge). The project could be any program to improve quality, such as improving the procedures by which computers are purchased for your company or improving the central filing system. One of the projects at the Paul Revere Insurance Company was to insure that every phone was always answered within three rings. Prompt attention was important to their customers. The initial project should be substantive and important for clients and the company, yet feasible. It should focus on system issues (chronic problems, not fire fighting) and provide the opportunity to use a heterogenous team—that is, a problem or opportunity that involves people from all levels of the organization. It should be readily amenable to establishing key measurements, preferably with baseline (pre-change) data available. The needed duration should be relatively short (four to six months), to avoid delaying the bigger program. Of course, the amount of time needed by the team members to participate should be clearly designated and available. When potential projects are considered, ask such questions as to what output is unsatisfactory, how do we know, what data do we have, what might the future data look like if the output is improved, and does it represent a good opportunity for improvement?

For the pilot and future projects, the steps involved in the team efforts include:

- Steering committee establishes a team sponsor. It would probably be the manager of the department or group most impacted by the project. The sponsor does not participate in the team meetings, but sees that any needed resources are made available. The sponsor will also work with the team in implementing their results.
- Steering committee and sponsor select the TQM team leader from the department or group that is most likely to be affected by changes recommended from the project.
- Steering committee, sponsor, and team leader select the team members (eight to

ten maximum) from all groups affected by the project, to create the opportunity to break down barriers between groups and create synergy between groups. Keep in mind that the initial project team members can become missionaries for the process throughout the organization. Look for people that welcome change, have a history of success, are open to training in interpersonal skills, and are willing to commit to the team for the duration of the project.

- Steering committee and sponsor define the team's project and purpose and guidelines for conducting the project in writing. This would include items such as schedule, budget, limits of authority, and procedures for revising any guidelines.

- Team meets to review project purpose and guidelines to see if scope, schedule, and budget are reasonable, and if the team members selected can effectively conduct the project. Team then makes recommendations on project revisions and team changes to steering committee and sponsor for approval.

- Steering committee appoints a liaison between team, sponsor, and steering committee. Liaison makes regular contact with team, acts as facilitator when needed, but does not run the meetings.

- Team establishes meeting frequency, format, groundrules, and notekeeping responsibilities. Notes are kept for future reference and learning by new teams. Ways to create an environment that is supportive to the team and its purpose are developed. This could include holding special meetings with all staff, to communicate the status of the project and get input and buy-in from the staff.

- Team establishes flow chart of process or system being studied in project and identifies problem areas. If system seems too big for the project, team selects a section of it that could be demonstrably impacted and improved in the time allotted for the project.

- Using the brainstorming and Pareto techniques discussed earlier, the team determines the key measures and operating definitions that would be used to show impact of changes on system from the project. Team begins to collect data, to establish baseline of key measures.

- While data are being collected, team uses the brainstorming, cause and effect diagrams, Pareto process, and "why" process to establish the root causes for the major problems identified in the system. This is then followed by using the same basic tools to develop implementable solutions to the root causes of the problems as described earlier.

- Team statistically evaluates the baseline data and prepares run charts, histograms, scatter plots, control charts, and so on. They then determine which key measures are likely to be affected by the proposed solutions, and if new measures will need to be monitored.

- Team presents recommendations of solutions to sponsor for review and approval. Recommendations include specific action steps and/or policy changes and suggested key staff who should be implementers of the changes.

- Sponsor initiates and/or delegates implementation of solutions. Team assists with implementation to the level of authority bestowed on them and continues to collect data on key measures.
- Team statistically evaluates data, to determine the impacts of the solutions and changes on the system. Then they prepare a project report and present it to the sponsor and steering committee.
- Team and sponsor continue to monitor results and implement other related solutions.

Guidelines for TQM Team Liaisons

The TQM team liaison has the vital task of following the team's progress and assisting them to keep on track and schedule.

- Check with the team leader weekly to see if they are on track. This may be done by attending the team meeting and talking to them afterwards or by calling them. Ask them the following questions:
  - How are you doing?
  - What problems are you having?
  - What will you do next?
  - How can we help you?
- If the team is off track, attempting a bigger project than originally scoped, or grinding to a halt, attendance at one or more of their meetings as an external facilitator is needed. The key in facilitating the meeting is not to jump in and try to fix things, but to:
  - Encourage the team to keep their goal or issue that they are trying to resolve visible to the group throughout the meetings. Refocus the group when they get off track.
  - Ask open-ended questions to find out where they are headed.
  - Clarify everyone's understanding of the issue.
  - Summarize the group discussion as you heard it.
  - Encourage silent members to participate by asking for their input.
  - Wield a Pareto cleaver when necessary, to cut off extraneous issues and efforts.
  - Allow the group to make the final decisions themselves. Don't let them push it onto you, as a management representative. It is their process.
  - If you have ideas, offer them as suggestions for consideration. Don't state an answer as obvious just to move the process along. If you find yourself getting drawn into the issue and have a personal attachment to a particular outcome, you are not acting as a facilitator but as a team member. This is a common pitfall. It is difficult to stay detached when you have a vested interest in the outcome.
  - Make suggestions to increase or decrease the frequency of the meetings,

depending upon which stage they are in. For example, with brainstorming, it would probably be more effective to have a longer session or two sessions close together when everyone is on a roll.

- When asked for suggestions, try to give two or more options for the team to consider, rather than just the answer you favor.

- During the meetings, observe the team leader. The team leader's role is to direct the team through the steps. It is tempting for the team leader to rush through steps without getting everyone's input, or to censor suggestions that may inhibit team members. It may be appropriate, if this is happening frequently, to ask open-ended questions of the group or by summarizing the different viewpoints in the room. Give the team leader specific feedback after the meeting on how to solicit everyone's input without censoring. Encourage the team leader to delegate tasks, such as subteam leadership, tasks to be done before the next meeting, data collection, and so forth.

- Praise the team leader and team at the end of the meeting for their progress, willingness, and skills demonstrated. A few good and sincere words go a long way.

- Encourage the team members to participate if you perceive them to be sitting back and expecting the leader to do all the work. The team leader may burn out by the end of the project and leave with a feeling of satisfaction but no desire to participate again, and the team will not be empowered.

- The team should have an agenda for every meeting, with times allocated for items. Assist them in developing groundrules for conducting the meetings to which they all agree.

- If the team is on track, check in weekly and occasionally attend meetings to show concern and nurturing for the team.

- Check in with the sponsors regularly, to discuss their concerns and perceptions of the issue and team progress. They are accountable, but have little control over the team except for setting the initial guidelines. In their first exposure to TQM, this will probably not be comfortable for them, moving from a state of control to empowering the employees. It is therefore important to nurture them too, by listening to them and sharing your perspective of the team's progress. This will be easier for the liaison than the team members, who may be influenced by other role relationships. An alternative is to be present when the team leader gives the sponsor a project update. Your role is to facilitate the communication and understanding of the issues and concerns of both the team and the sponsor. Assist the team leader in effectively communicating the team's progress with the sponsor and the staff.

- Assist the team and the sponsor in determining the points at which updates should be given to the staff. It is probably a good idea to show the staff charting of the data being collected, say every two weeks. This helps maintain everyone's interest in the project, reducing the resistance to the time spent collecting data, and building buy-in to the solutions by keeping everyone informed and involved. The teams may be so involved in the process that they forget to keep everyone informed. Maintaining the right environment is critical to success.

- The team may need to create subteams to deal with certain aspects of the project concurrently (e.g., determining the validity of existing data, developing a draft of data collection forms, and preliminary analyzing or checking of new data). The subteams will make recommendations to the team.
- Encourage teams to communicate with each other directly and/or through you. The teams can learn from each other and save each other from some pitfalls.
- Assist the teams in determining the ways in which the data will be analyzed for key measures evaluation.
- Let all the team members know that you are available as a resource to help them succeed.
- Remember that the team process is iterative, not linear. The team will cycle through three stages—getting started, working together, and reaching agreement—at least twice. Keep them moving, keep them on track and within scope, and nurture and appreciate the team.

Guidelines for TQM Team Leader

Following are guidelines for the TQM team leader:

- You are the contact point for communication between the TQM team, sponsor, liaison, steering committee, and others outside the team.
- Develop meeting agendas; schedule and conduct meetings. Assist team in developing groundrules for conducting meetings to which they all agree, such as:
  - Raise your hand to talk, wait until called upon by team leader. Don't interrupt.
  - In case of disagreement, brainstorming of options will be conducted, followed by Pareto voting, or the liaison will be called in for assistance.
  - If you know you will miss two meetings, designate a replacement (consistent). If you miss more than two meetings, you may be replaced.
- Act as a full participating team member.
- Keep all team records, including correspondence, meeting minutes, charts, graphs, and other data related to the team project.
- Work with the TQM liaison to provide training in statistical methods to the team and use the liason as a resource to address other issues.
- Share responsibility for the team's progress and success by encouraging team members to participate as a recorder, data gatherer, flipchart writer, idea presenter, or subteam member.

- Build and maintain full participation by all team members through effective and frequent use of questions to draw out all team members' ideas and contributions.
- Create subteams to deal with certain aspects of the project concurrently, such as evaluating existing data or drafting data collection forms.

Guidelines for TQM Team Members

Some guidelines for TQM team members include:

- Actively participate in all team activities by attending all meetings and by sharing experience and knowledge of the system under evaluation.
- Your suggestions and opinions are equally as important as other team members' or the team leader's, regardless of the roles or titles of the team members. By the same token, listen to, respect, and consider other team members' opinions and suggestions.
- Gather, prioritize, and analyze data. This includes working between team meetings, to gather relevant information and forward the TQM team's efforts.
  - Interviewing staff and customers.
  - Observing processes.
  - Gathering data.
  - Charting data.
  - Writing reports.
- Assist team leader in conducting meetings, volunteer for appropriate subteams, and share in the workload.
- Help team prepare progress reports and final results for presentation to steering committee.
- Help develop and follow through on action plans and solutions being implemented by sponsor and team.
- Use the liaison as a resource.
- Share responsibility for the team's progress and success.

Guidelines for the TQM Team Sponsor

Following are some guidelines for the TQM team sponsor:

- Establish clear guidelines to the team on their limits of authority, budget, schedule, and status reporting and review dates.
- Provide support and resources to the team to ensure their continued progress and success.
- You will need to walk a fine line between showing concern and support for the team and not appearing to be pushing for control or interfering in the team's progress. This will be trial and error to a degree.

- Schedule separate review or status meetings with the team or team leader. Do not attend regular team meetings.
- Demonstrate your support of the team to all your staff in all communications. Attend all status meeting for the entire staff to show your support.
- In review meetings with the team:
  - Review progress since last meeting.
  - Solicit feedback on problems team has encountered.
  - Inquire how you can help them with resources, training, and so forth.
  - Find out the team's next steps and determine if they are consistent with project purpose and scope.
- The liaison is available as a resource to you and the team if the project grinds to a halt or if there is a problem or conflict.

Problems Teams Typically Encounter

Some problems that teams will encounter include the following:

Problem: The scope of the project expands so that one or both of two results occur:
  - The project begins to impact other groups, sections, and departments where there is no buy-in, or
  - The project cannot be completed within the agreed-upon time frame.

Response: The liaison and sponsor must wield a Pareto cleaver on the scope of the project. The parts that get eliminated can be put on a list of future projects for TQM teams.

Problem: Meetings go off on tangents, or items are discussed to death and nothing gets done.

Response: The team leader must establish and stick to an agenda that includes items to be covered, decisions to be made, time limits for items, and person responsible for item. The liaison can be called in to facilitate a meeting, if this is still a problem, to demonstrate how to run the meeting.

Problem: Some team members may not participate or are not as comfortable speaking up, while others are fast to take the floor.

Response: The team can establish a groundrule that each person must raise a hand and be acknowledged by the leader before they can speak. Brainstorming techniques, where everyone writes down ideas and expresses one at a time in turn, will also work to draw out quiet members.

Problem: A team member stops attending meetings or regularly misses them.

Response: First, review the groundrules to see if absenteeism is covered. Consider replacing them, or, if it is a temporary absenteeism, they should designate a substitute who is qualified to present the point of view of the group that the person represents.

Problem: The meetings are full of conflict or deadlocked, or there appear to be cliques forming on the team.

Response: The liaison should be called in to facilitate the meetings as an impartial party. The team will tend to open up to the neutral party. The liaison should use communication skills to clarify the points of view and brainstorm alternative ways to handle them. The liaison will need training in conflict resolution.

Problem: The rest of the staff don't buy into the project and resent collecting data, because of the extra effort required.

Response: The sponsor needs to stress support for the team to the staff. The team needs to communicate the progress and the potential benefits of the project and of collecting data to the staff by holding more status meetings. At these meetings, the staff's input should be solicited, to develop buy-in. Appreciation should be expressed to all staff for their participation and extra work.

Problem: Management or the sponsor does not like the recommendations from the TQM team.

Response: The steering committee and liaison should be involved, to find out the sponsor's concerns and determine if it is a policy/procedure or personal control issue that needs to be addressed. The team and the sponsor may not be able to resolve the issue alone because of the role relationships. Encourage the sponsor to take a "what if" attitude. What if this recommendation resulted in an improvement instead of the chaos you are speculating will happen? Do not discount practical concerns about the implementation; brainstorm ways to resolve them. The steering committee must provide support and assurance to the sponsor in the case of success or failure of an implemented change.

Publicize the Results of the Initial Projects

I broke my nose five times, lost one-third of my teeth in football. Do you think I'd do that if there were no fans in the crowd and no scoreboard at the end of the field?

Thomas Malone
President, Milliken & Company
1989 Winner, Malcolm Baldrige Award

Distribute the good news from the initial project to the staff. This is also an opportunity to reinforce the company commitment to the quality program. A monthly newsletter on the quality program that publicizes the quality things being done by coworkers has been effective. It can be a vehicle for spreading the details of improvements, as well as a vehicle for recognition and celebration of a team's efforts.

Implement the Full Program

The training of the trainers should be completed and training for all personnel initiated. Quality teams incorporating everyone in the company are formed, as training is completed. The quality program is publicized throughout the company. Keys in the early implementation phases involve the building of communication, trust, education, experimentation, involvement of everyone, and recognition of the successes.

As the program begins, the teams will likely be a mixture of enthusiasts and cynics. Some will be convinced that this is another program of the month that will soon fade into oblivion. Experience has shown that even the most cynical will be affected by watching fellow workers coming up with ideas that improve their working environment and performance. Once the cynics decide there is some merit to the process, they are already in a position as a team member to contribute. Make it easy for everyone to contribute!

Experiences to date with programs in service industries are more limited than manufacturing, but offer some hints on quality team procedures:

- *Everyone* in the organization is eventually on a quality team.
- Keep team sizes at eight to ten people per team.
- Quality team leaders should be rotated annually, to bring new ideas and life to the team.
- No one should be on more than one team.
- Some cross-departmental teams have merit. Quality is a great way to pull divergent elements of an organization together into a team.
- Dedicate a specific amount of time to the team—one-half hour per week or one hour every other week have been successful.

- Provide personal expressions of gratitude by management for successes, as well as broader recognition programs.

Continuous monitoring of the results and feedback to achieve further improvement is an integral part of the full program.

Within this brief presentation, we've introduced concepts of a quality improvement program. The references provide added information that will be useful to you as you pursue a program for your own company. You've no doubt noticed that many of the basic skills that we've discussed in the first eight chapters of this book come into play. A quality program provides a way to structure the application of these basic skills in conjunction with formal quality techniques developed in other fields. Although the application of these formalized quality procedures to service industries has less of a history than in manufacturing industries, the following benefits have resulted in service industries:

- For individuals:
  - Ideas are heard.
  - The sense of identity with the company is improved.
  - Breeds creativity and innovation; individuals grow and learn.
  - A sense of contribution.
  - Fun.
  - People are less resistant to change.
  - Ideas under the quality program carry more weight, and people feel a sense of commitment because they have really helped in the decision-making process.
  - People like to do a good job; it makes them feel good about themselves.
- For the company:
  - Increased profit and productivity.
  - The right people are making changes—the people who do the work.
  - Improved morale.
  - Top management is regarded more positively because employees are trusted to formulate and implement their own recommendations. It eases the "us" versus "them" attitude.
  - Staff turnover is reduced.
  - Less litigation on projects.
  - Client satisfaction improves, which is what it's all about!

References
1. Simon, R.C. 1990. Total quality management in design and construction. *The Construction Specifier*, p. 102. May 1990.
2. Saarinen, A.W., and M.A. Hobel. Setting and meeting requirements for quality. Amer-

ican Society of Civil Engineers. *Journal of Management in Engineering*, p. 177. April 1990.

3. Scherkenbach, W.W. 1986. *The Deming Route to Quality and Productivity.* ASQC Quality Press.

4. 1990. *Development of a Company-Wide Quality and Productivity Improvement Process.* QualPro Company.

5. Baker, B.N., D.C. Murphy, and D. Fisher. 1988. Factors affecting project success. *Project Management Handbook* (2nd edition), edited by D.I. Cleland and W.R. King, p. 902. New York: Van Nostrand Reinhold.

6. Crosby, P.B. 1979. *Quality is Free: The Art of Making Quality Certain.* New York: New American Library.

7. Heidenreich, J.L. 1988. Quality program management in project management. *Project Management Handbook,* edited by D.I. Cleland and W.R. King, p. 513. New York: Van Nostrand Reinhold.

8. Randolph, W.A., and B.Z. Posner. 1988. *Effective Project Planning and Management—Getting the Job Done.* Englewood Cliffs, New Jersey: Prentice Hall.

9. Garfield, C. 1987. *Peak Performers—The New Heroes of American Business.* Avon.

10. Gitlow, H.S. 1990. *Planning for Quality, Productivity and Competitive Position.* Homewood, Illinois: Dow Jones-Irwin.

11. Townsend, P.L., and J.E. Gebhardt. 1990. *Commit to Quality.* New York: John Wiley and Sons.

12. Deming, W.E. 1986. *Out of Crises.* Cambridge, Massachusetts: MIT, Center for Advanced Engineering Study.

13. Hulnick, H.R. 1989. "One Way to Know," unpublished course material, University of Santa Monica, California.

14. Pascarella, P., and A. Fruhman. 1989. *The Purpose-Driven Organization.* San Francisco, California: Josey-Bass.

15. Levinson, H. 1990. Appraisal of what performance. *People: Managing Your Most Important Asset.* Harvard Business Review.

16. Zemke, R. 1991. Bashing the Baldrige. *Training,* p. 29. February 1991.

17. Orsburn, J.D., L. Moran, E. Mussselwhite, and J.H. Zenger. 1990. *Self Directed Work Teams—The New American Challenge.* Homewood, Illinois: Business One Irwin.

18. Sensenbrenner, J. 1991. Quality comes to city hall. *Harvard Business Review,* p. 64. March–April 1991.

10

Exercises

There is no substitute for first-hand experiential learning through training exercises and games. They provide the opportunity to practice new skills, perhaps that the participants have only read about, in a safe, comfortable environment. Everyone is in the same boat, so the process of learning is less threatening and more fun. Exercises allow and encourage the less expressive group members to participate fully in this safe environment.

Group exercises take the focus away from the leader or trainer and puts it on the individuals in the group. They become responsible for their own learning. Each participant will get something different out of the exercise. They will remember what is important and relevant to them in their current life or work situations. Learning comes from being an active participant and from being an observer of other participants in a group process.

Every time an exercise is conducted, the training and results vary because of the different participants involved. Treat each exercise as a life situation, even if it appears contrived. During the exercise, be aware of your own behavior and your reactions to others' behavior. Evaluate the effectiveness of your behavior and reactions, and consider if changes could be made to become more effective. Remember that the only behavior you have control over is your own!

The exercises in this chapter have been designed to be conducted by anyone who has an interest in learning, for themselves and others. A background in training and facilitation will help, but is not necessary. All exercises are accompanied with instructions and handouts, all of which may be copied and used. They can all be conducted as stand-alone sessions (e.g., as brown bag lunches for staff or for project teams). References have been made to chapters in this book in each exercise. It is suggested that the leader and the participants, if possible, read the relevant chapters before conducting the session.

There are twelve exercises in this chapter that could constitute a three-month training program of weekly training sessions.

IMPROMPTU PRESENTATIONS AND USE OF TRANSITIONS

Suggested Reading: Chapter 2

Purpose: To give participants practice in giving spontaneous presentations and in developing transitions from one topic to the next, with minimal preparation time.

Time Required: Approximately 5 minutes per participant.

Materials Required: At least two 3 x 5 cards per participant. Eggtimer.

Preparation: Prepare a series of 3 x 5 cards, each having a word on it. Use a mixture of technical and nontechnical topics, to make the transitions from one topic to another more interesting. Use technical topics that are applicable to your particular business.

Examples:

Technical	Nontechnical
Project	Grapefruit
Schedule	Monkey
Budget	Mountain
Task	Feelings
Deadline	Headlights
Process	Keys
Goal	Radio
Client	Shoes
Communication	Picture
Resources	Sailing
Team	Balance
Misunderstandings	Light

Procedure: Tell the participants that you will pull a card from the deck and call on a participant at random. That participant must then stand and immediately give a 60-second spontaneous talk on the topic on the card. Time the talk with an egg timer that the participant can see. After 60 seconds, pull another card and call on another participant. The next participant must stand up and immediately use a one- or two-sentence transition from the previous topic to the next, and then give a 60-second spontaneous talk on the new topic.

Example: Transition from Deadline to Headlights.

Meeting a deadline is a lot like seeing a deer on the road in front of you in your headlights—you only have a finite amount of time to take action before missing or "hitting" your deadline.

This procedure is followed for each of the participants in turn. Each participant should get at least two turns.

Optional. Videotape all the presentations and view the tape at a second training session, to observe presentation style.

EXCHANGE OF IDEAS AND USE OF PARETO PROCESS TO REACH CONSENSUS

Suggested Reading: Chapters 3 and 9

Purpose: To encourage participation and sharing of ideas between participants and to demonstrate process for reaching group consensus.

Time Required: Approximately 30 to 45 minutes, depending on the number of participants.

Materials Required: Three 3 x 5 cards for each participant, flipchart pad, easel, and colored markers. Prizes for the inventor and nominator of the top three ideas. Box to hold cards.

Preparation: Give each participant three 3 x 5 cards the day before the session. Tell them to write one positive idea or suggestion on each card, based on a theme or issue that you choose.

Examples:

1. How to improve communication between engineering and drafting, or research and development, or production and marketing.
2. How to avoid duplication of work effort.
3. How to estimate hours for a task more accurately.
4. How to inspire someone to be self-motivated.
5. How to monitor project progress relative to budgeting more effectively.
6. How to solicit and use input from all team members.

Let them know that the ideas will be shared and considered by the group, to develop consensus on the best ideas. The participants should bring the cards to the session and sign their names on each card.

Procedure: Ask the participants to make sure their names are on their cards. Collect the cards, shuffle them, and put them into a box. Each participant draws three cards (not his or her own), reviews them, selects the best one, in their opinion, and signs their own name *below* that of the idea's inventor. Collect all the selected best idea cards, and write the ideas on the flipchart (do not identify the inventor or nominator). Take a few minutes to allow the group to clarify any ideas that are unclear or confusing. Participants should not identify themselves as the idea's inventor during this discussion period. Each participant then votes on the ideas using the 10/4 voting method described in Chapter 9 (each person has 10 votes and the maximum number of votes they can expend on one idea is 4. So they could vote 4,4,2; 4,2,2,2; 4,3,3; 2,2,2,1,1; etc.) After the voting, the top three ideas with the most votes qualify for prizes, which are awarded to the inventor and the nominator. The rest of the group wins by gaining new ideas. After the session, summarize and distribute the ideas to the group for their use.

HANDLING OBJECTIONS

Suggested Reading: Chapters 2 and 4

Purpose: To enable participants to practice responding to objections quickly, effectively, and with ease.

Time Required: Approximately 10 minutes per participant.

Materials Required: Egg timer or stopwatch, 3 x 5 cards. A pen and pad for each participant to record good ideas and responses.

Preparation: Prepare a list of possible objections typically encountered in the marketing, presentation, or implementation of projects. Transfer each objection onto a 3 x 5 card.

Examples:

1. Your estimate is 30 percent higher than it was a few months ago.
2. Your fees are higher than firm B's.
3. Why are your estimated fees for the project so much lower than all the others?
4. We only have a limited budget to get this project done and it is less than your estimate. What are you going to do about it?
5. We don't pay travel expenses.
6. Your staff appears inexperienced.
7. Your team appears to be too big.
8. Your scope is not clear in Task A.
9. Your PM appears to be overcommitted.
10. This project will cause major problems for the residents in the area.
11. The environment will be irreparably damaged by your project.
12. These alternatives have not been evaluated thoroughly enough.
13. The public won't understand this technical jargon.
14. The board won't like the alternative and its cost.
15. This project will not meet our deadline.
16. Nobody will buy the cost sharing recommended in your project.
17. This is too risky an alternative, politically.
18. We would rather use a local firm.

19. Your schedule is unrealistic.

20. Your report isn't what I expected. You need to rewrite it.

21. Surely you can do better than that.

These questions need to be customized specifically to your business. If they are too general, the participants may either get confused and stuck or give an answer that is too vague.

Procedure: Shuffle the deck of 3 x 5 cards. One participant draws a card and reads the objection out loud. The participant sitting to the left of the participant drawing the card must respond to the objection. The participant must respond immediately and spontaneously. Allow one minute for the response, which can be timed using a stopwatch or egg timer. Stop the participant at the time limit. This will encourage preciseness and conciseness. Then allow a couple of minutes for the rest of the group to make suggestions for additional responses.

Encourage participants to write down responses that they want to remember and use in the future. Then the participant who just finished responding to the objection draws a new card for the next person on the left. This is repeated until each participant has had at least two turns at responding.

TV TALK SHOW INTERVIEW

Suggested Reading: Chapter 2

Purpose: To become aware of the frequency of use of closed and open-ended questions, and to practice using open-ended questions more.

Time Required: Approximately 45 minutes.

Materials Required: Copies of instructions.

Preparation: Make enough copies of instruction sheets A, B, and C for a third of the participants. Make a copy of instruction sheet D for each participant. Provide a pen and paper for each participant.

Procedure: Divide the group into subgroups of three participants (trios) by a random counting method. Ask them to arrange their chairs facing the direction as shown. Draw the arrangement on a board or flipchart to show them. All trios should be in this arrangement.

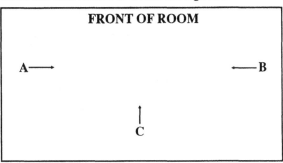

Tell them that they will be conducting TV talk show interviews in their groups, and each participant will get a turn to be interviewee (A), interviewer (B), and observer (C). Give them a minute to decide which role they want to try first, move their chairs, and sit in their designated chair.

Hand out the instruction sheets to the participants as follows:

Sheet A to interviewee
Sheet B to interviewer
Sheet C to observer
Sheet D to each participant

Tell the participants to read their sheet(s). Give A one minute to decide the topic that they want to be interviewed on. Then give B three minutes to think of open-ended questions to ask A in the interview. Remind the groups that TV talk show hosts want to get as much detail out of their guests as possible, so they should avoid closed questions that can be answered with a yes/no or other one-word response. Ask the groups if they have any questions before starting the interview. Allow five minutes for the actual interview. After the interview, tell the observer, C, that they have five minutes to tell B how many closed and open-ended questions were asked. They can also give B feedback on the questions that were particularly effective, and make suggestions for additional open-ended questions. Then rotate the roles counter-clockwise, so the observer in chair C moves to chair B and becomes the interviewer; the interviewer in chair B moves to chair A and becomes the interviewee A; and the interviewee in chair A moves to chair C and becomes the observer. The instruction sheets A, B, and C should remain on the same chairs for each rotation. Repeat the interview process until everyone has played all three roles.

If there is time, the process can be repeated, to see if the number of open-ended questions increase and the closed questions decrease.

TV Talk Show
Instruction Sheet A

Person Being Interviewed

You have one minute to think of a topic you are interested in and are willing to be asked questions about (e.g., a favorite hobby, sport, or some strongly held views about politics, environment, how you would change the world etc.)

When you have decided, tell B, who is going to interview you for the TV talk show, what the topic is.

B will then have three minutes to think of questions to ask you. You may want to use this time to think of answers to potential questions.

You will be interviewed when the session leader tells you to start.

TV Talk Show
Instruction Sheet B

The Interviewer

You are a TV talk show interviewer who is to interview A about a topic of A's choice. A has one minute to decide on the topic.

After A has told you the topic, you have three minutes to plan your questions. You should try to use open-ended questions that produce more than a yes/no or one-word answer. These questions often start with the words "how, why, where, what, when," and so on. You should avoid closed questions that produce only yes/no or one-word answers as these give you very little time to think of your next question.

When you are interviewing, listen to what A says and try to produce questions that follow on as naturally as possible. This will show that you are listening and following A and not just asking questions from a prepared list.

Interview A when the session leader tells you to start.

TV Talk Show
Instruction Sheet C

Observer

The other two members of your trio are going to act out a TV talk show. The interviewee, A, will have one minute to think of a topic and the interviewer, B, will have three minutes to prepare questions.

When the interview starts, you are to count the number of open-ended and closed questions that the interviewer B asks and mark them on instruction sheet D. Record briefly the questions asked that were particularly effective and make notes of other questions that you think B could have asked.

At the end of the interview, you will be asked to tell B how many closed and open-ended questions were asked. You will also give B feedback on questions that were very effective and give suggestions for additional open-ended questions.

It is vital that you concentrate and stay aware of the interview, even though you will not be speaking. Make enough brief notes to help you remember the questions, but not too many, or you will miss part of the interview or next question.

TV Talk Show
Instruction Sheet D

	Use Check Marks to Count Questions (IIII)
Closed Questions	
Open-Ended Questions	

Questions that were very effective:

1
.
2.

3.

4.

5.

6.

7.

8.

Suggestions for additional questions:

1.

2.

3.

4.

5.

6.

7.

8.

ONE-WAY COMMUNICATION

Suggested Reading: Chapter 2

Purpose: To demonstrate the impacts on comprehension when only one-way communication is allowed and ambiguous words or instructions are used.

Time Required: Approximately 10 minutes.

Materials Required: Several sheets of square paper at least 8-1/2" x 8-1/2".

Procedure: Ask the participants if their instructions or communication with someone (a team member, client, etc.) have ever been misinterpreted. Tell them that this exercise explores how misinterpretation might occur. Ask for four volunteers to stand in front of the group. Give them each a piece of paper, and tell them that they must close their eyes and not ask questions for the duration of the exercise. Give them the following instructions. Watch their response and give them enough time to complete each step before giving them the next step.

Step 1. Fold the paper in half and then tear off the bottom right corner of the paper.

Step 2. Fold the paper in half again and tear off the upper left corner.

Step 3. Fold the paper diagonally corner to corner and tear off the bottom corner of the paper.

Tell them to open their eyes, unfold the paper and show it to the participants. (It is highly unlikely that all four will be the same.)

Ask the participants (observers and volunteers):

1. What was ambiguous about the instructions, based on their own observations and interpretations.

2. How could the instructions be clarified or the process be improved?

This exercise could be followed, in the same session, by the exercise on one- and two-way communication.

ONE- AND TWO-WAY COMMUNICATION

Suggested Reading: Chapter 2

Purpose: To demonstrate the pitfalls of one-way communication and the benefits of effective two-way communication, especially on phone calls.

Time Required: Approximately 40 minutes.

Materials Required: Each participant will need a pencil, eraser, and two sheets of blank paper. An overhead projector and screen will be required in the room.

Preparation: Make one copy of each of the two diagrams on the following pages. Make one overhead slide of each of the diagrams. Arrange the room so that there is one chair at the front, facing away from the group. The group should be seated at tables where they can draw.

Procedure: *Exercise 1—One-Way Communication*

Ask for a volunteer to assist in the exercise. Tell the group that the volunteer will be verbally describing something to them. They are to follow the instructions and sketch the diagram or object being described.

Give the volunteer diagram #1. Instruct the volunteer to sit in the chair facing away from the group and give the group verbal instructions only (no gestures) on how to draw the diagram. The group is not allowed to ask questions. Each participant will draw the diagram based on the instructions heard. Note the time it takes to compete the exercise. At the completion, show the group the correct diagram on the overhead projector. Ask participants to compare results and show the group their diagrams. Ask the group:

How could there be so many interpretations of the same instructions?

Exercise 2—Two-Way Communication
Give the same volunteer diagram #2. Instruct the volunteer to turn the chair and sit facing the group for the second exercise. The volunteer will again give verbal instructions on how to draw the diagram. However, this time the participants are allowed to ask

questions. The volunteer can clarify instructions, but not show the diagram to the group. Note the time it takes to complete the exercise. At the completion, show the group the correct diagram on the overhead projector. Ask the participants to compare results and show the group their diagrams. Ask the group:

1. What conclusions could they draw from the two exercises?
2. Guess how long it took to complete each exercise. (Then give them the actual times.)
3. Was the extra time taken in the second exercise reflected in the accuracy of the drawings?
4. How could two-way communication be improved to result in higher accuracy?

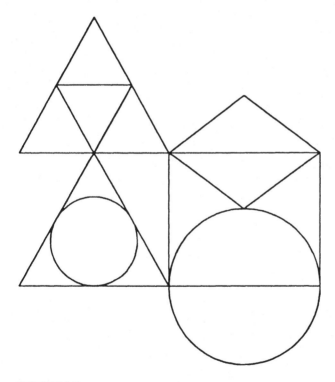

DIAGRAM #1

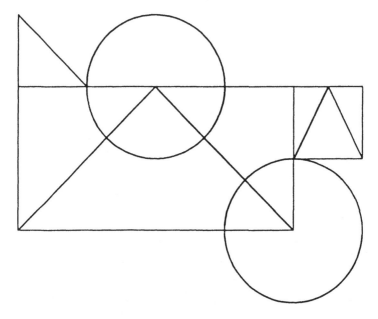

DIAGRAM #2

REDUCING TIME WASTERS

Suggested Reading: Chapters 5 and 9

Purpose: To explore ways to be more personally effective through identifying activities that waste time and brainstorming solutions to reduce or control the time wasters.

Time Required: Approximately 45 to 60 minutes.

Materials Required: Flipchart, overhead projector and screen, easel, and colored marker pens.

Preparation: Make a copy of the list (adopted from Jim Richardson of TIME Inc.) of time wasters for each participant. Make an overhead sheet of the list.

Procedure: Hand out the list of "Time Wasters" to all participants. Tell them that they have five minutes to check all the time wasters that currently impact their effectiveness, whether caused by themselves or others. Then tell them to rank the top five time wasters out of both columns at the bottom of the page (1 being the biggest time waster).

Place the "Time Wasters" list on the overhead projector. Get each participant to read out loud their top five time wasters. Mark them on the overhead. After all results are received, determine the combined top five time wasters for the group, based on those most frequently in the top five for individual participants.

Then ask the participants to brainstorm ideas for reducing or controlling the group's top five time wasters. They will write them down in silence for the next five minutes. Then go around the group, getting one idea from each participant in turn and write them on the flipchart. Repeat the rounds until everyone has run out of ideas. Participants will say "pass" if they don't have another new idea.

Tell the group to write down five of these ideas that they are willing to implement for themselves over the next month. Tell them to pick a partner in the group who they will check in with once a week. The purpose of checking in is to share successes and get assistance with problems in implementing the ideas.

Optional. Schedule another session in one month's time. At that session, give all the participants, in turn, five minutes each to talk about the successes and problems they've experienced in implementing the ideas. Write the successes and problems on two flipcharts.

When the lists are complete, ask the participants for suggestions on how to solve some of the problems as a group. Allow 20 minutes for discussion of solutions. Encourage the participants to again write down the suggestions that they want to implement.

TIME WASTERS

Caused by Self

_____ 1. Inadequate planning
_____ 2. Preoccupation
_____ 3. Ineffective delegation
_____ 4. Attempting too much
_____ 5. Too involved in details
_____ 6. Reverse delegation
_____ 7. Unable to say no
_____ 8. Arguing
_____ 9. Socializing
_____ 10. Fear of offending
_____ 11. Unrealistic time estimates
_____ 12. Unable to terminate visits
_____ 13. Failure to anticipate
_____ 14. Goals not clearly defined
_____ 15. Emotional upset
_____ 16. Personal disorganization
_____ 17. Fatigue
_____ 18. Low morale
_____ 19. Poor filing system
_____ 20. Outside demands
_____ 21. Shifting priorities
_____ 22. Procrastination
_____ 23. Interruptions
_____ 24. Mistakes
_____ 25. Failure to listen
_____ 26. Lack of monitoring
_____ 27. Poor communication
_____ 28. Unable to control meetings
_____ 29. Others?_____

Caused by Others

_____ 1. Overlong visit
_____ 2. Waiting for decision
_____ 3. Unnecessary meetings
_____ 4. Delays
_____ 5. Interruptions
_____ 6. Multiple bosses
_____ 7. Duplication of effort
_____ 8. Unclear objectives
_____ 9. Lack of information
_____ 10. Trash mail
_____ 11. Office noise/talking
_____ 12. Management by crisis
_____ 13. Responsibility without authority
_____ 14. Poor communication
_____ 15. Negative attitude
_____ 16. Mistakes
_____ 17. Lack policies
_____ 18. Lack authority
_____ 19. Understaffed
_____ 20. Overstaffed
_____ 21. Socializing
_____ 22. Lack of feedback
_____ 23. Role not clear
_____ 24. Low-priority memos
_____ 25. Shifting or changing priorities
_____ 26. Lack of clerical staff
_____ 27. Interruptions (visitors)
_____ 28. Paperwork
_____ 29. Others?_____

Top Five Time Wasters

_____ 1._____

_____ 2._____

_____ 3._____

_____ 4._____

_____ 5._____

HANDLING LOADED QUESTIONS

Suggested Reading: Chapter 2

Purpose: To give participants practice, in a safe environment, at handling loaded questions that have an underlying emotion or concern.

Time Required: Approximately 10 minutes per participant.

Materials Required: Egg timer or stopwatch, 3 x 5 cards.

Preparation: Prepare a list of possible loaded questions that are typically encountered in the marketing, presentation, or implementation of the projects. Identify the emotion or concern underlying the question.

Examples:

1. I live near this new highway and the noise keeps me awake. What makes you think the sound barriers work? (Anger)
2. How can we be sure that you will actually be the person that works on the project and not disappear as soon as we award it to you? (Distrust)
3. I don't believe your project approach "A" will work. I've done a very similar one using approach "B" that worked very effectively. How would you compare the two? (Ego)
4. Explain your project to me again. Are you saying that Alternative A or B is the most effective for us? (Confusion)
5. Are you telling me that I now have to go back to my Board and tell them that the project that I guaranteed would be within budget is now going to be overbudget by 20 percent? (Fear)
6. This is news to me. You're telling me the project is undersized because of errors. Why didn't you catch it earlier? What are you going to do about it? (Anger)

Transfer each question onto a 3 x 5 card. Put the concern at the bottom of the card in parentheses.

Procedure: The first participant to answer the question stands in
front of the group. Another participant selects a card
and reads the question out loud. It is important that the
participant reading the question conveys the emotion
or concern listed on the card in their tone or expression.
The participant in front of the group gets two minutes
to address the concern or emotion and answer the
question. (Note: They are not necessarily related!)
Stick to the two-minute limit using a stopwatch or egg
timer.

At the end of the two minutes, ask the participant
answering the question to identify the concern or emotion underlying the question and describe how it was
addressed in the answer. (Example: I perceive the person to not *trust* me.) Ask the questioner if they felt that
the concern or emotion had been adequately addressed.
The rest of the group can make suggestions for additional responses. Participants should be encouraged to
make notes on responses that they want to remember
for future reference.

Repeat the process until each participant has responded
to at least two loaded questions.

OBSERVING NONVERBAL BEHAVIOR

Suggested Reading: Chapters 2 and 4

Purpose: To become more aware of body language and nonverbal communication, and practice asking open and closed questions.

Time Required: 5 minutes per participant and 20 minutes of discussion at the end.

Materials Required: Copies of instructions.

Preparation: Make enough copies of instruction sheets A and B for half of the number of participants in the group, and copies of instruction sheet C for all participants. Put one chair at the front of the room where the session is to be held, away from any other furniture, so the body language of the person sitting in it can be fully observed.

Procedure: Shuffle instruction sheets A and B, and hand each participant one of the sheets. Tell them not to show them to anyone else. Hand out instruction sheet C to everyone.

Tell the group that each participant will take a turn in the "hot seat" and will speak for two minutes following the instructions on their sheet.

Tell the group that some participants will be telling the truth and that some will not. The purpose of the exercise is to determine who is and who isn't.

After each participant has spoken for two minutes, the group has the opportunity to ask open-ended (what, when, where, how, who) questions and closed (yes/no) questions for three minutes to find out more information and observe nonverbal reactions. Each participant writes observations on instruction sheet C, but keeps silent about their conclusions.

After all the participants have spoken and been questioned, the group discusses each participant in turn. The group discusses impressions of who was telling the truth and who was not, and reasons for the impression based on nonverbal communication, as well as verbal information. Allow 15 to 20 minutes for the discussion. Finally, ask the participants who were not telling the truth to identify themselves.

At the conclusion of the session, ask the participants to observe people's nonverbal behavior for the rest of the day and be aware of their own interpretation of the message being sent. The message may be one of discomfort, lack of commitment, lack of confidence, excitement, enthusiasm, urgency, impatience, and so forth.

Truth or Deception
Instruction Sheet A

You will be asked to sit in the "hot seat" to talk about yourself for two minutes. **You should always tell the truth.**

You might like to talk about your life history, your family, your interests, or your previous work experience.

After you have spoken, you will be asked questions by the group. Answer these as honestly as possible.

Answer closed questions with Yes or No only. Answer open-ended questions with additional information.

When you are watching other people in the "hot seat" that you know well, and know they are or are not giving accurate information about themselves, keep it to yourself. Watch their nonverbal communication anyway.

Truth or Deception
Instruction Sheet B

You will be asked to talk about yourself for two minutes.

You should not tell the truth.

You might like to invent your life history, your family, your interests, or previous work experience.

After you have spoken, you will be asked questions by the group. Try to make them believe your story.

Answer closed questions with Yes or No only. Answer open-ended questions with additional information.

When you are watching other people in the "hot seat" that you know well, and know they are or are not giving accurate information about themselves, keep it to yourself. Watch their nonverbal communication anyway.

Truth or Deception
Instruction Sheet C

Write down your observations on each person as you listen and watch them talk.

Participant Name	Notes on Observed Nonverbal Behavior Watch for eye contact, eye movements, anxiety, blushing, nervous movements, a calm air, squirming, gestures, hand movements, open/closed body position, facial expression, congruence in verbal and nonverbal expression.	Telling the Truth? Yes—Y No—N

EXPLORING DREAMS AND DEFINING GOALS

Suggested Reading: Chapters 3 and 5

Purpose: To turn dreams into goals by exploring the experience desired if the dream is fulfilled.

Time Required: Approximately 45 minutes.

Materials Required: Copies of "Ideal Scene" and "Goal" sheets, flipchart, easel, and colored markers.

Preparation: Make copies of the "Ideal Scene" and "Goal" sheets for each participant.

Procedure: Hand out the "Ideal Scene" sheets to the participants. Tell them to imagine they could be or do anything they wanted. They should put aside their limiting beliefs that make them see their dream as impossible, such as no time, money, experience, talent, and so on. The only criteria is to *want* to be or do, rather than a "should" placed by themselves or others on them. Tell them to write this dream down in the center of the "Ideal Scene" sheet. Assure them that they can keep it secret if they want to. Tell them to write on each of the "Ideal Scene" spokes what it would look like or the experience they would have if they realized their dreams. These are not action steps towards reaching the dream, but are descriptions of what it would be like when it is achieved. Write the following example on the flipchart before the session to illustrate the process. Give them 15 minutes to complete the "Ideal Scene" sheet.

Example:

Dream: Run and own a business
(in center)
Desired
Experience: Have more control of my life.
(on spokes)   Be able to work on projects I choose.
                   More self-satisfaction and
                     accomplishment.
                   Take time off to be with family.
                   Making at least $150,000 a year.
                   Eager to go to work everyday.

Then hand out the "Goal" sheets and ask the group to write down a goal that could be achieved in the next three months, that would realize some of the experience desired from the dream. For example: In three months be able to choose the projects I want to work on, or take more time off with the family. It does not necessarily mean achieving the dream or moving towards it by quitting a job. It will probably mean taking the steps to produce the experience in the current situation.

The next step is to break the goal down into three realistic, achievable, and manageable monthly steps. Write the following example on another flipchart sheet before the session, to demonstrate the process.

Example:

Three Month Goal—To be able to choose 60 percent of the projects I work on, and to take three weeks off with my family, knowing my projects are being handled well.

First Month Action Step—Research upcoming projects to decide which I want to work on. Find out what experience and time commitment is needed from people working on project. Arrange a meeting with supervisors to let them know what I want to do and ask them what needs to be done by me or them to arrange it.

Second Month Action Step—Make a list of all tasks I do that I don't have anyone trained to handle if I was away. Identify team member(s) to substitute for me. Meet with them to discuss what time or training is necessary for them to substitute for me. Plan the training over the next two months. Check back with supervisors to reaffirm interest in projects.

Third Month Action Step—Give people I work with my vacation schedule six weeks in advance. Ask them to start planning around the time accordingly. Make sure that deadlines are not scheduled within three days of the vacation. Renegotiate ahead of time if there are conflicts. Ask supervisors for a commitment on my participation in the project.

Give the participants 15 minutes to define the goal and action steps. When they have completed the "Goal" sheets, suggest that they transfer their steps into their planners or "to-do" lists to remind them to take action. Get each participant to select a partner with whom they will check in on a monthly basis to discuss successes and solve problems.

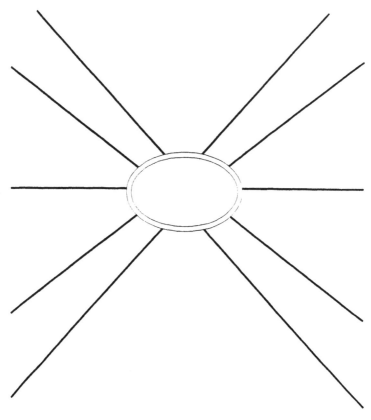

Creating an Ideal Scene

DEFINING A GOAL

My three-month goal to move toward achieving the experience desired from realizing my dream is:

My first monthly action step to move towards my three-month goal is to:

My second monthly action step to move towards my three-month goal is to:

My third monthly action step to move towards my three-month goal is to:

STATEMENTS OF FACTS VERSUS OPINIONS

Suggested Reading: Chapters 2 and 6

Purpose: To illustrate the difference between facts or data and opinions or assumptions.

Time Required: Approximately 20 minutes.

Materials Required: Two ordinary looking objects. Flipchart, easel, and color markers.

Preparation: Visit a novelty or magic store and buy two objects that appear normal, but are not what they seem (e.g., rubber pens, artificial food, trick cigarettes).

Procedure: Hold up one of the objects. Ask the participants in turn to make one statement of fact or observation about the object. Mark these on the flipchart. Go around the group until everyone runs out of statements. Participants say "Pass" when they run out of statements. Then ask the group to decide whether each statement is really a statement of fact or just an opinion or assumption about the object. Mark an "F" next to the facts and an "O" next to opinions. Count the numbers of Fs and Os. Then show the group that the object is not really what it seems to emphasize the difference between statements of facts and opinions.
Ask the group the following questions:

1. What difference would it make in meetings or discussions if everyone clarified whether statements made were opinions or facts? *(Less chance of miscommunication or actions based on individual or group opinion.)*
2. What constitutes a statement of an opinion? *(Statement goes beyond what was actually observed. Assumptions are made about the topic or object.)*
3. What constitutes a statement of fact or data? *(Statement limited to what was actually observed directly by the observer.)*

Hold up a second object and repeat the exercise. There will probably be discussion as the statements are made as to whether it is fact or opinion. Encourage the discussion. Count the number of Fs and Os again, to show the level of improvement once participants become aware of the difference between statements of facts or opinions.

PROJECT SCHEDULING

Suggested Reading: Chapters 6, 7 and 8

Purpose: To practice developing schedules, to practice getting input from team members on project planning, and to practice group planning meetings.

Time Required: Approximately 80 minutes for a group of 18. Add 5 minutes for each additional 6 people.

Materials Required: One copy of the workplan, guidelines, and the project schedule sheet for each participant, and one set of contractor and subcontractor information sheets for each team. Overhead projector and screen in the room. One overhead slide of the project schedule sheet and one overhead pen for each team. A prize for the shortest workable schedule developed by a team.

Procedure: *Part 1*

Divide the participants into teams, 6 people per team. Tell the participants that they are contractors selected to build a house and that each team is to develop a construction schedule. Hand out the guidelines, assignment, work plan, and project schedule sheets to each participant. The information for the contractor and each subcontractor is to be copied onto separate sheets of paper. Distribute one set of these sheets to each team. Each team member is to take one sheet at random. Explain that the separate sheets represent the common situation where each team member has information on the timing of his or her own task, but needs input from other team members on the relative timing of their tasks. It is the purpose of the meeting to get input from all team members so that a coordinated schedule is prepared. Suggest that each group establish ground-rules for their team meeting and appoint a chairperson, facilitator, recorder, and timekeeper. Tell them that they have 50 minutes to develop a schedule on their own. After 50 minutes, ask if anyone needs more time. If so, allow up to 5 more minutes. Allow 5 more minutes for each team to transfer their schedule to the overhead slide of the schedule sheet.

Part 2

Ask each group to nominate a representative. Each representative will then, in turn, show their schedule on the overhead projector to the group and explain the scheduling selected.

Keep track of the total schedule time in days for each group, to determine the shortest schedule.

Ask the participants:

1. Did the group with the shortest schedule follow the guidelines? If not, go to the second shortest schedule, and so on, to determine the winner.
2. Did the group follow the groundrules they agreed upon at the start of the exercise?
3. What problems did they encounter in their meetings? How might they be addressed?
4. What were the benefits of getting input from all group members?
5. How could the scheduling of projects in the office be improved?

Award the prize to the group with the shortest schedule that follows the guidelines.

GUIDELINES

You are the general contractor and subcontractors selected to build a house. Each of you have prepared an estimate of how long your task will take (attached on the Work Plan Sheet). The general contractor has a signed contract with the owner and now wants to schedule the minimum construction time. The contract has an incentive clause. The team will receive $300 for every day less than 120 days that it takes to finish the house.

You are to use the sequence of construction steps described on the information sheets given to the general contractor and each subcontractor. Although each subcontractor knows where their work falls in sequence, you must hold a team meeting to determine how to schedule all tasks to minimize the construction time. In addition to sequencing the tasks, also consider logical timing (e.g., don't complete cabinets before they can be installed, to avoid unnecessary storage and to avoid incurring costs that cannot be reimbursed until the cabinets are installed).

ASSIGNMENT

Prepare a schedule, based on the guidelines, the workplan, and the contractor and subcontractor information sheets, showing the start and finish times for each task. Be prepared to discuss your reasoning for the schedule.

INFORMATION SHEETS

Put the information for each contractor and subcontractor on a separate sheet of paper.

General Contractor

As the general contractor, in addition to obtaining the permits and coordinating subcontractors, you will perform the following tasks. Each of your tasks can immediately follow other tasks as shown below:

Carpets, install—after all other interior work.
Fireplace—after insulation.
Landscaping—immediately before final inspection.
Sheetrock—after insulation.
Site preparation, Grading—after permits.
Water/sewer hookups—after permits.

Carpentry Subcontractor

You will perform the following tasks, which can immediately follow other tasks as shown below:

Cabinets, build—after owner selections complete.
Cabinets, install—after sheetrock.
Decking, exterior—after roofing.
Finish work—after interior painting.
Framing, siding—after foundation.
Roofing—after framing, siding.
Windows, install—after framing, siding.
Wood flooring—after sheetrock is installed in areas with wood floors.

Concrete Subcontractor

You will perform the following tasks, which can immediately follow other tasks as shown below:

Foundation—after site preparation, grading.
Driveway, sidewalks—immediately before final inspection.

Electrical and Heating Subcontractor

You will perform the following tasks, which can immediately follow other tasks as shown below:

Electrical wiring installation—after plumbing.
Electrical hookup to utility—after permits.
Electrical fixture installation—after interior painting.
Heating, air conditioning—after electrical wiring installation.
Insulation—after heating, air conditioning.

Painting Subcontractor

You will perform the following tasks, which can immediately follow other tasks as shown below:

Painting (exterior)—immediately prior to interior painting.
Painting (interior)—after sheetrock.

Plumbing Subcontractor

You will perform the following tasks, which can immediately follow other tasks as shown below:

Plumbing—after framing.
Plumbing fixtures (install)—after cabinets installed.
Tiling—after sheetrock is installed in area with tile.

PROJECT SCHEDULE
TIME (Days)

Task	7	14	21	28	35	42	49	56	63	70	77	84	91	98	105	112	119	126	133

WORKPLAN

Tasks (Alphabetical Order)	Duration (Days)
Cabinets (build)	28
Cabinets (install)	4
Carpets, install	4
Driveway/sidewalks	2
Electrical wiring installation	7
Electrical hookup to utility	1
Electrical fixture installation	2
Exterior decking	10
Final inspection	1
Fireplace	6
Finish work (woodwork, hang doors, etc.)	8
Foundation	5
Framing, siding	20
Heating, air conditioning	8
Insulation	5
Landscaping	6
Owner selects:	
-Plumbing fixtures	2
-Electrical fixtures	2
-Carpet	2
-Paint	3
Painting (exterior)	4
Painting (interior)	4
Permits	14
Plumbing	10
Plumbing fixtures (install)	4
Roofing	5
Sheetrock	14
Shipping (carpet, fixtures)	28
Site preparation/grading	4
Tiling	5
Water/sewer hookups	4
Windows, install	5
Wood flooring	6

References

1. Christopher, E.M., and L.E. Smith. 1987. *Leadership Training Through Gaming.* London: Kogan Page, New York: Nichols Publishing Company.
2. Bond, T. 1986. *Games for Social and Life Skills.* New York: Nichols Publishing Company.
3. Newstrom, J.W., and E.E. Scannell. 1980. *Games Trainers Play.* New York: McGraw-Hill Book Company.
4. Scannell, E.E., and J.W. Newstrom. 1983. *More Games Trainers Play.* New York: McGraw-Hill Book Company.
5. Scholtes, P.R. 1989. *The Team Handbook.* Madison, Wisconsin: Joiner and Associates, Inc.

Index

Lightning Source UK Ltd.
Milton Keynes UK
UKOW04n1644201017
311330UK00001B/17/P